*Advance Praise for*

## PLIGHT OF THE LIVING DEAD

"This book is fantastic! The sci-fi stories you've read barely hold a candle to the gruesome ways in which parasites manipulate their hosts in real life. This book will make your skin crawl with some of the best examples of manipulation we've encountered, fascinate you with what we know about how parasites achieve these amazing feats of control, and leave you wondering what this all means for the nature of free will. You'll be thinking about this book long after you're done reading it."

—Kelly Weinersmith, *New York Times* bestselling co-author of *Soonish: Ten Emerging Technologies That'll Improve and/or Ruin Everything*

"Matt Simon is, to borrow his term, a zombifier: *Plight of the Living Dead* will infect your brain, forcing you to spout a stream of bizarre facts—about fat-sucking worms, muscle-eating fungi, brain-stabbing wasps—until your friends buy the book for themselves, and the chain of infection continues."

—Mark Essig, author of *Lesser Beasts: A Snout-to-Tail History of the Humble Pig*

"A gruesome, fascinating, and somehow hilarious exploration of the most devious, mind-altering tactics of the bug wars. I found myself cringing, laughing, learning, but most of all thankful I'm not an ant."

—Cody Cassidy, author of *And Then You're Dead: What Really Happens If You Get Swallowed By a Whale, Are Shot from a Cannon, or Go Barreling over Niagra*

PENGUIN BOOKS

# PLIGHT OF THE LIVING DEAD

**Matt Simon** is a science writer at *Wired* magazine, where he specializes in zoology, particularly of the bizarre variety, and the author of *The Wasp That Brainwashed the Caterpillar*. He is one of just a handful of humans to witness the fabled mating ritual of the axolotl salamander. He lives in San Francisco.

ALSO BY MATT SIMON

*The Wasp That Brainwashed the Caterpillar*

# PLIGHT of the LIVING DEAD

***What the Animal Kingdom's Real-Life Zombies Reveal about Nature—and Ourselves***

MATT SIMON

PENGUIN BOOKS

PENGUIN BOOKS
An imprint of Penguin Random House LLC
375 Hudson Street
New York, New York 10014
penguinrandomhouse.com



Quotations in Chapter 8 from *The Conquest of a Bombus terrestris Colony by a Psithyrus vestalis Female* by Cor Van Honk et al., *Apidologie*, 1981, 12 (1), pp. 57–67.

LIBRARY OF CONGRESS CATALOGING-IN-PUBLICATION DATA
Names: Simon, Matt, author.
Title: Plight of the living dead: what the animal kingdom's real-life zombies reveal about nature—and ourselves / Matt Simon.
Description: New York, New York: Penguin Books, [2018] | Includes bibliographical references.
Identifiers: LCCN 2018004526 (print) | LCCN 2018015907 (ebook) | ISBN 9781524705145 (ebook) | ISBN 9780143131410 (paperback) |
Subjects: LCSH: Parasites. | Predation (Biology)
Classification: LCC QL757 (ebook) | LCC QL757 .S48 2018 (print) | DDC 591.6/5—dc23
LC record available at https://lccn.loc.gov/2018004526

Printed in the United States of America
1 3 5 7 9 10 8 6 4 2

Set in Utopia Std | Designed by Sabrina Bowers

*For all those humans out there who've had the common decency not to rise from their graves.*

# Contents

# Introduction

***Welcome to the surreal world of the real-life living dead.***

It's a mythical creature that's so familiar, it may as well be real. The human gone wrong, the shell of a person racked with a virus that, in retrospect, really took scientists by surprise. Symptoms include stiffened joints and the consequent outstretched arms. The moaning, of course. Sunken eyes. Sometimes the beast just kind of stands there, as if lost in thought. Bits of the creature are falling off—toes and such, which won't be missed. And let's not forget the yearning for human flesh and general refusal to die.

The zombie—or living dead, or walking dead, or undead, or really any dead other than *dead* dead—is a monster phenomenon. You were probably a zombie for Halloween once. At the very least you've taken too much NyQuil and *felt* like a zombie. Hollywood puts out so many zombie films every year, something like 95 percent of Americans have acted in at least one of them.* There's zombie comedies and zombie romances and even zombie reimaginings of classic literature because yeah sure why not. Our culture is obsessed with the zombie, a legend that both fascinates us and forces us to confront tricky questions about what it means to be human.

The zombie may as well be real not just because Hollywood can't seem to quit it, but because the thing makes biological sense. Not the reanimating corpses bit—that's unreasonable. But think about it from the theoretical virus's perspective: It has to find new hosts, and what better way to do that than to assume control over the zombie's mind, making its victim yearn for

---

*Give or take a few percentage points.

human flesh? One bite and *boom*, transmission. A virus makes its way around a population not by way of sneezes, but through sophisticated behavioral manipulations that turn its host into an unwitting vehicle.

The zombie may as well be real because it actually is, only in a far more incredible and diabolical and horrifying way than a screenwriter could ever dream up. Because all across the animal kingdom, parasites are climbing into other creatures and mind-controlling them. Be they worms or wasps or microbes, certain organisms have figured out how to brainwash their victims in ways so clever and precise, they make Hollywood's creations look downright irresponsible.

In September 2013, I was pacing in my kitchen, talking on the phone with presumably a madman. In South America, he told me, a fungus invades ants' bodies and takes over their minds, manipulating them with unreal precision and consistency. The parasite steers the ants out of the colony and up a tree always at noon, always ordering them to bite onto a leaf always about a foot off the ground. This just so happens to be where the temperature and humidity are ideal for the fungus's growth. And the body snatcher has positioned its host right above the colony's trail, so as it erupts out of the back of the zombie ant's head and sprays its spores, it infects more victims. A parasite without a brain of its own has brainwashed one of the most loyal creatures on Earth to betray its family in spectacular fashion.

This mad scientist, the world expert on the zombie ant, was putting me in a tough spot. As he walked me through the manipulations and the mechanisms and the horrifying implications, I walked myself around the room with the chills, excited as hell to write about his work, but almost certain that he was conning me. My job as a science writer was and is to listen to people who dedicate their lives to science, then tell their stories in ways that make sense to humans. But pacing that kitchen—which slanted dramatically to the south, by the way, so I had to concentrate on the most bizarre thing I've ever heard *and* on not tipping into the fridge—I couldn't help but think I was about to broadcast the ravings of an overactive imagination.

But the fungus is very real, just one of a stunning number of species that have evolved powers of mind control. And I don't say that lightly. I mean very specifically *mind control.* I mean parasites that not only invade the bodies of their hosts, as parasites are wont to do, but that in precise ways hijack the minds of their hosts. Like the so-called brainworms, which lodge in other animals' heads and steer them around. Some bacteria turn their victims into zombies, too. Viruses do it as well. Even a certain barnacle, of all things, invades a crab's body and grows through its tissues like tree roots to command it around the ocean depths. Like the zombie virus of fiction that infiltrates humans, twisting them into the kind of hyperaggression that's liable to get the parasite passed on to its next host, the mind controllers of reality bend their victims to their will.*

Which seems fantastical and, well, impossible. But there's a perfectly reasonable explanation for all of this. A parasite hacking its host's mind is a matter of hacking pure biology, not some nebulous notion of consciousness or a soul. It's about flooding the brain with neurotransmitters like serotonin or dopamine,

---

*To put a definition on it: A traditional zombie is an undead human whose body has been hijacked by a pathogen, typically a virus, which manipulates the creature into spreading said pathogen by way of aggression, particularly biting. The many parasites we'll meet in this book are zombifiers, then, in that they assume control of their hosts' minds and sometimes physically alter their hosts' bodies to benefit themselves.

glitching how the victim perceives its world. Or about hypnotizing the dupe with sights or smells. All of it, no matter how unbelievable, is grounded in physiology that scientists are just beginning to understand.

Whatever their methods, the zombifiers have profound implications for how we think about the animal kingdom broadly, and parasitism specifically. Because more than half of the animal species on this planet are parasites. That puts you and me in the minority. And then there are all the bacteria and fungi and viruses outside the animal kingdom that exploit other organisms. With an estimated 9 million total species roaming this world, and scientists having described a little over a million of those, there must be far, far more brainwashers of which science is ignorant. That's partly a problem of this being a big planet with lots of nooks and crannies to explore, but also a problem of our own prejudices. We're visual creatures, so the parasitic manipulations we can appreciate are usually the ones we can see. Who knows how many parasites manipulate the sounds their hosts make, or the smells they exude. Undiscovered zombifiers are surely capable of exploits that make our ant-invading fungi look sloppy. Now it's just a matter of finding them.

I want you to come with me on a journey through the strange science of parasitic mind control. We'll meet the perfectly sane man who keeps zombie ants in his lab. And we'll wander into the forests of New Mexico in search of the crickets whose zombifying worms command them to leap into water and to their potential death. And don't forget the wasp that performs brain surgery on cockroaches. They're all real, and years after I first learned that it's possible for one organism to assume total control over the mind of another, the zombies are still making me pace my kitchen in a different apartment, this one not slanted dramatically to the south.

This is going to get weird. Because our journey isn't just about the zombifiers per se, but also what they reveal about nature—and ourselves. To truly understand the surreal world of the mind controllers, we'll have to leave our human prejudices behind. We'll have to better appreciate how the zombie sees and feels and smells its environment, how the parasite hijacks the senses of the victim to turn the body against itself. And, I'm sorry to say, we'll have to abandon the delusion of free will, for humans or zombies or otherwise. Because the zombifiers prove better than anything on this planet that the mind is a material thing, and is therefore easily hacked. And the human brain, I'm also sorry to say, is no exception: Parasites can twist our minds as brutally as the fungi do to ants and the wasps do to cockroaches, for we aren't nearly as special as we think we are.

# 1

# The First Rule of Zombification: You Do Not Want to Be a Zombie

***Wasps—aka the flying middle fingers of the animal kingdom—prove that evolution is the meanest and most beautiful thing the world has ever known.***

Nestled in a basement of Israel's Ben-Gurion University is Frederic Libersat's room of nightmares, a tiny box filled with still tinier boxes of cockroaches and the most conniving insect on Earth: *Ampulex compressa,* aka the emerald wasp or jewel wasp. Its hypnotizing beauty belies its belligerence, with big eyes and a precious green sheen to its body, colors shifting as the parasite sprints about the cage.

Today is an unlucky day for this roach. It scrambles around a petri dish at the rear of the cage and seems to realize the mistake it's made. It freezes and rights its posture, like a tiny lowrider, but the iridescent wasp pounces anyway, grasping the roach and slamming her stinger between its front legs, paralyzing them. As the pair grapple in a Looney Toons–esque cloud of limbs and wings, the wasp, who's half the size of the cockroach, pulls out her stinger and—now unobstructed by normally flailing front legs—drives it through her victim's neck and into the brain. They keep tussling, flinging the cage's lovely pastel pebbles of blue and pink in all directions as the roach, desperate for leverage, braces itself against the wall of the plastic cage. Still the wasp, her body bent in half to make the reach, clings with her stinger stuck in the victim's head. But she's not dillydallying on account of pumping an ungodly amount of venom. No, she's feeling around for the right spot in the brain to inject her mind-control potion.

Two spots, to be exact. While the combatants flutter, the wasp's ultradexterous stinger probes for a pair of regions in the brain that govern locomotion. When she finds them, she loads each with venom, releases her grip, pulls out her stinger, and backs away. And the cockroach . . . stands there. After a few more minutes, instead of panicking like you or I might if a needle just went through our brains, the roach begins grooming itself, taking its legs and antennae into its mouth. Calmly, systematically, one limb after another, like nothing happened.

Ten minutes after the attack, the wasp again approaches. Without the slightest objection from the cockroach, she bites onto the base of its left antenna and slides her mouth a quarter of the way down, then snaps her jaws with a bob of her head. After a dozen tries—slide and bite, slide and bite—the wasp finally severs the antenna and laps blood from the stub. Then she does the same to the other—slide, slice, slurp. After all, brainwashing a cockroach takes a lot out of a wasp, and she needs to get back that lost energy.

Still the cockroach does not object.

When the wasp has had her fill, she again leaves the cockroach. But she returns a few minutes later, grabbing onto the left antenna stub and giving it a tug. Nothing doing—the roach won't budge. She leaves, returns, and tries again. The cockroach still won't follow. So she tries once more.

This time the potion has kicked in, and the dupe is ready. The wasp grabs the cockroach by the antenna stub and starts yanking it toward a vial in the middle of the cage. And the roach follows. It could walk just fine in the other direction if it wanted, mind you, but it accompanies its diminutive captor through the opening of the vial and into its tomb without complaint. Here the wasp lays an egg on the cockroach's belly, then leaves and rummages around for those pastel pebbles, piling them into the vial. One by one, carefully she chooses the right-sized rocks to get a good seal. And never once does the cockroach try to force its way out—not that the wasp is plugging up the entrance to keep her victim in there. Instead, she's doing it to keep the neighborhood opportunists from consuming her prisoner.

Inevitably the egg will hatch into a larva that pierces the roach's abdomen and sucks its blood. When that well runs dry, the young wasp will drill into the body and consume the organs, taking care to save the bits most vital for the roach's survival until the end—namely the central nervous system. After the tormentor has hollowed everything out, it metamorphoses, then erupts as an adult wasp from the cockroach's corpse, tunneling out of the tomb-turned-womb and into a world it will help make so cruel.

At the risk of starting this book out on a negative note, I must say that nature is a terrible place. Not for you and me and a lot of other humans, mind you. But for every other organism out there, life is suffering. It's starvation, it's disease, it's a creature bigger and meaner than you chewing on your head. Even apex predators like lions and bears have to worry about their own parasites and food shortages and the ravages of drought. There is no dying peacefully in sheets with high counts. And the zombifying wasps prove it better than any other animal on Earth: Nature is not fun, and it does not care about your feelings or well-being.

This doesn't bother me in a general kind of way. But watching the jewel wasp at work makes me sad. I have no love for cockroaches, believe you me. It's just how the parasite pounces and methodically drives its stinger into the roach's brain and *lingers*, with the victim's head bent in a silly way, staring at me through the plastic wall of the container. The cockroach doesn't speak English, and I don't speak Cockroach, but the idiot in me can't help but think it's asking for help. Which wouldn't do it any good because it's already too late. You know, the old cliché of a zombie biting your friend and you having to leave them behind because you know the virus is on its way to their brain.* Inevitably, the roach, too, will turn into a zombie. And not long after, a hollowed-out shell of an insect.

Such suffering troubled Charles Darwin. The Christian view of the world in the nineteenth century was a peaceful one, with animals living in harmony on an idyllic planet tailor-made by God. But what Darwin realized was that the animal kingdom is one giant scrum, a frenzy of death and suffering, of teeth and claws and stingers. And it was the wasps—specifically the ichneumonids, which lay their eggs inside other creatures such as caterpillars—that really troubled him. "I own that I cannot see, as plainly as others do, and as I should wish to do, evidence of design and beneficence on all sides of us," he wrote in 1860. "There seems to me too much misery in the world. I cannot persuade myself that a beneficent and omnipotent God would have designedly created the Ichneumonidæ with the express intention of their feeding within the living bodies of caterpillars, or that a cat should play with mice."

The problem for caterpillars and mice and our cockroaches is that cruelty begets cruelty. Now, let me be clear that "cruelty" is a human construct. No creature but a human can find a wolverine chewing on its leg and think, "Wow, this is cruel." So yes, when I say that nature is cruel, I'm being a judgmental human. But what the wasps do to cockroaches and caterpillars is so very

---

**Shaun of the Dead* put a good twist on this cliché, though you might argue that a normal human keeping his undead friend chained up in a shed so the two can play video games together ad infinitum is a selfish maneuver on the normal human's part, as it is assumed that a zombie would prefer rambling about town instead.

unpleasant because it's a tough world out there—no one is going to voluntarily babysit their young. Which means the victims are going to need some . . . convincing.

## This Might Sting a Little

What Frederic Libersat is teasing apart in his little room of nightmares is a sting like no other on Earth. A snake might inject a mouse with venom that paralyzes muscles, most importantly those that keep the lungs breathing air and the heart pumping blood. But the jewel wasp's potion is far more intricate. That first sting paralyzes the cockroach's front legs so it can't bat away the wasp's brain-sting, sure, but the second sting doesn't paralyze the roach at all. Instead it sends the creature into what's known as a hypokinetic (meaning "below normal movement") state. The dupe can full well move, but it can't *decide* to, even as the wasp gnaws off its antennae and drags it into a dungeon where a larva consumes it alive. Libersat knows it can move because if you throw the stung cockroach into water, the stress will jolt it awake, sending it scurrying across the surface like a normal roach would. Which means its muscles work fine.

"There's nothing wrong there," Libersat says. He looks like a slimmer version of Pablo Picasso, by the way, with shorn hair and eyebrows that bounce up and down his face when he gets excited. "It has really something to do with the animal engaging in walking, to make the decision that it has to move." With a head full of venom, and without the sudden stress of something like nearly drowning, it can only sit there for the wasp to torture.

Why the roach would do such a thing is a big, complicated question that Libersat is just beginning to answer. But clearly, figuring out how a wasp can sting a cockroach in the brain to convince it to live the rest of its miserable zombified life as food for a larva begins with decoding that venom, which is composed of over two hundred different compounds. To start with, Libersat has found that it's loaded with a substance called gamma-aminobutyric acid, or GABA, which you'll find across the animal kingdom. This shuts down the front legs in that initial sting, but only for about five minutes. "GABA is a known inhibitory neurotransmitter in the central nervous system," Libersat says. "Basically what the wasp is doing is using existing machinery that produces this neurotransmitter." Meaning, the wasp didn't evolve some novel compound to cripple the cockroach—it weaponized the GABA that the roach and wasp share to short-circuit the muscles that power the front legs.

When it comes to the brain sting, things get a bit trickier, considering this is, well, brain surgery. Why the wasp would want to force the cockroach to groom itself obsessively following the sting, or if there's any significance to it at all and it isn't just a bizarre side effect, is still a puzzle. But *how* the wasp makes it happen seems clear: dopamine. This compound is most famously associated in humans with pleasure—it's one of the reasons why cocaine is so expensive—but serves several other functions, including the regulation of grooming. Bump up dopamine levels in rats and you get more grooming, for instance. Humans, too, as it happens. "A drug addict has problems with dopamine," Libersat says, "because they take drugs that increase the level of dopamine, showing very often a stereotypic behavior of scratching, which is equivalent to grooming." Unfortunately for the cockroach, the wasp delivers a dose of dopamine-loaded venom not into the bloodstream, but straight to the central nervous system, leading to that same kind of obsessive grooming. Whether that serves to distract the cockroach, or if it's a way for the wasp to keep its host nice and clean and parasite-free (save for its own murderous larva, of course), or if, again, the behavior serves no benefit, isn't clear.* But you can probably blame dopamine for it.

Another tantalizing behavior of a stung cockroach is the way it carries itself, a phenomenon that Stav Emanuel, a postdoc in Libersat's lab, has been decoding. "They have decreased posture," Emanuel says. "The posture is weaker and their head is pointed down and their antennae are limp." It's the same kind of posturing you find in a sleeping roach. Not only that, but when Emanuel hooked up the leg muscles of stung and slumbering cockroaches to an electromyograph, or EMG, she found the electrical activity to be eerily similar between the two groups compared to their awake, unstung counterparts.

So the wasp may have pulled off a brilliant evolutionary maneuver here. Much like it probably weaponized GABA—an inhibitory neurotransmitter it didn't need to bother inventing—to shut down the front legs, by stinging the cockroach in the right part of the brain the wasp may have also hijacked a preexisting state: sleep. Beyond the posture, a stung roach acts like a sleeping roach in that it's far less sensitive to stimuli, which may explain why it lets the wasp chew off its antennae and push it around. So by lulling the cockroach with a sting to the brain and erasing its self-preservation, the wasp can walk its much larger

---

*Now's as good a time as any to define what a parasite actually is. That'd be an organism that lives on another organism (a tick on your leg, for instance) or within another organism (a tapeworm in your gut) at the expense of the organism, known as a host. Typically that means the parasite is extracting energy from the victim in the form of nutrients.

Parasites can have their own parasites, by the way, known as hyperparasites. Oh, and a parasite that eventually kills its host, like the jewel wasp larva does, is called a parasitoid. And a parasitoid that targets parasitoids is called a hyperparasitoid, which must look fantastic on a business card.

victim into the lair instead of having to drag it there. Here the roach sits dazed, unable to escape the larva that devours it alive. It's all reminiscent of the sleep paralysis you or I might suffer: You wake up and you're conscious, but your body is frozen. Only the cockroach's problem is it'll never snap out of it—that, and a larva is consuming it alive.

Now, brain surgeons have their drugs, but they also have their instruments, and the jewel wasp's stinger is one magnificent tool. Any wasp or bee that stings you is a female, for that stinger is a modified ovipositor, which the insects use to lay eggs. (More dramatically put: The stinger makes life and takes life.) But unlike the brute force of a honeybee stinger—designed to drive into flesh and pump as much painful venom as possible—the jewel wasp's weapon is far more precise, a rapier to the honeybee's claymore sword.* It's loaded with sensors that let the wasp feel its way as it penetrates the cockroach's neck and brain sheath and finally into those specific bits of the brain. Indeed, remove a cockroach's brain and give a jewel wasp the corpse and it'll feel around for ten times longer for the right spot to inject venom. Dip

---

*In case you were curious, the jewel wasp's sting for a human feels like barely anything at all, just the prick of the tiniest of needles. The jab of the wasp known as the tarantula hawk, on the other hand, may well be the most painful sting on Earth. At two inches long, this colossal wasp operates much like the jewel, though it skips the convoluted mind control and outright paralyzes tarantulas, dragging them into its den as living food for its young.

Tangle with one of these wasps and you'll regret life itself. The sting is full-tilt electric, delivering pain so intense that in an actual scientific paper, an actual scientist recommended just lying down and screaming until the misery passes. (Mercifully, the pain lasts only three minutes.) Otherwise, you'll end up sprinting around in a panic and hurting yourself. One researcher who once took a series of stings in a row "crawled into a ditch and just bawled his eyes out," because he was a professional, goddamnit.

a wasp's stinger in liquid nitrogen and the manipulator gets even more flustered, feeling around for an average of twenty minutes compared to the typical sixty seconds. (Libersat did this by coaxing the wasp to sting through a film, like you would to milk venom from a snake or spider. But instead of extracting venom, he dripped liquid nitrogen on the wasp's needle.) If she can't feel her stinger, she can't feel the brain.

So the wasp has stung its cockroach, crammed it into a den, laid an egg, and sealed it up. In this tomb the zombified roach loses its instinct for self-preservation, but a subtler manipulation is unraveling inside its body. Because unlike a snake, which immediately kills and consumes its prey, the mother wasp has to keep her victim unspoiled for a week, enough time for her larva to drain the life out of it and transform into an adult. Outright paralyzing the cockroach is a no-no—that'd mean it couldn't breathe. "No gas exchange means that the meat is going to rot quicker," says Libersat. So not only does the jewel wasp's venom keep the victim breathing, but it slows down the roach's metabolism. Meaning Mom injects a preservative of sorts to keep the roach plumper and fresher for her youngster.

By turning a roach into a living larder, momma jewel wasp has evolved one hell of a way to help her kids survive. But why such cruelty? Why plunge a cockroach into such a nightmare? To answer that, we'll need to first explore the greatest idea that biology has ever known.

## A Brief and Dare I Say Fascinating Story of the Theory of Evolution by Natural Selection

Stuck in bed with a nasty fever in 1858, Alfred Russel Wallace did what many of us do while stuck in bed with a nasty fever: He changed science forever.

All right, maybe it wasn't just a fever. It was probably malaria—he was laid up in the Malay Archipelago, after all. And maybe Wallace wasn't an ordinary human. He was a brilliant young naturalist exploring the jungles of Southeast Asia, nabbing beetles and birds and butterflies and anything else that moved for shipment back to England, where a burgeoning community of biologists consumed specimens—only intellectually, of course—with a full-tilt mania.

Suffering there on the island of Halmahera, Wallace ruminated on any old idea that popped into his head, included among them Thomas Malthus's famous hypothesis that things like famine and disease have a habit of keeping the human race from growing too numerous. Logically, Wallace might apply this to the entire animal kingdom. That, of course, would suggest "enormous and constant destruction." And so Wallace wondered: Why do some individuals of a given species live while others die nasty deaths?

Then came the realization, as he recounted in his autobiography:

> And the answer was clearly, that on the whole the best fitted live. From the effects of disease the most healthy escaped; from enemies, the strongest, the swiftest, or the most cunning; from famine, the best hunters or those with the best digestion; and so on. Then it suddenly flashed upon me that this self-acting process would necessarily improve the race,

> because in every generation the inferior would inevitably be killed off and the superior would remain—that is, the fittest would survive.

Wallace had fever-dreamed the idea of evolution by natural selection, a little more than a year before Charles Darwin published *On the Origin of Species*. And speaking of . . . The next thing Wallace did was mail his idea, scrawled on twenty pages of rice paper, to Charles Darwin.

Unbeknownst to Wallace, Darwin had for over two decades been grappling with the same realization, which he knew would blow science's head right off its shoulders. Darwin understood well what his idea would do to the established religious order in England, not to mention that it'd upend the tenets of biology, so he took his time amassing evidence for it. After all, he'd figured out how species could come to be without a higher power.

So when Wallace's manuscript reached Darwin on a June morning in 1858, he was—how should we say—a bit upset. Later that day he wrote his friend, the geologist Charles Lyell, with whom he'd shared his theory of natural selection, and enclosed the manuscript. "It seems to me well worth reading," he somewhat understated, adding: "Please return me the M.S. which he does not say he wishes me to publish; but I shall of course at once write and offer to send to any Journal. So all my originality, whatever it may amount to, will be smashed." Yet even without permission from Wallace—it would have taken months to get a letter to him, and still more months to get a reply—Darwin's gentleman-scientist friends decided the proper course of action would be to present the findings of both men to a meeting of the Linnean Society. (Back in those days, you had to formally submit papers by reading them to a room of Generally Old White Men.)

By all accounts, the reading of the Darwin-Wallace joint paper at Burlington House in Piccadilly by one George Busk was 100 percent unremarkable. No one screamed about the heresy, no one fainted, and no one started a fistfight. It wasn't until Darwin published *On the Origin of Species* the next year that things really kicked off—when the backlash from God-fearing scientists, and even-more-God-fearing religious figures, took full effect.

But what of Wallace? Considering you may have never heard of him, history sort of kicked him in the shins on this one. Yet his contributions not only to natural history—of wandering around collecting specimens of never-before-described creatures—but to evolutionary theory have been mighty, beyond him spurring Darwin into publishing the theory of natural selection. (Wallace pretty much invented the field of biogeography, the study of how and why species are distributed on Earth.) And all the while, he was laid-back about Darwin's compatriots sharing Wallace's work without his permission. In fact, he was honored. In a letter to the botanist Joseph Hooker, one of the coconspirators, he wrote: "Allow me in the first place sincerely to thank yourself and Sir Charles Lyell for your kind offices on this occasion, and to assure you of the gratification afforded me both by the course you have pursued, and the favourable opinions of my essay which you have so kindly expressed."

And so the theory of evolution by natural selection came to be. A century and a half later, it remains the foundation on which biology stands. And without it, we cannot begin to comprehend how zombification in the animal kingdom could happen without the guiding hand of a Generally Old White Man in the Kingdom of Heaven.

What Darwin and Wallace had independently realized is that life is a drag. Modern humans have a reasonable likelihood of dying of old age, but not so for everything else in nature. Life is suffering. And it's this suffering that drives the evolution of manipulation.

Here's the reality. Organisms produce more offspring than can possibly survive. For mammals like ourselves, that means as insurance we might pump out a few more than the two kids that would replace their mother and father in a stable population, because previous to the invention of modern medicine, the odds were pretty much zero that all offspring would survive to themselves reproduce. In any population, many individuals won't make it to adulthood, and even then, they still have to grapple with any number of their own predators and diseases. Matters get all the more apocalyptic when you look at fish and insects, which are capable of pumping out not dozens or hundreds or thousands of eggs in a go, but *hundreds of thousands*. Think of the odds there if on average two siblings survive to replace their parents in the population. The cold cruelty of nature will claim the rest, be it by way of predation or starvation or natural disaster.

It's this cruelty that powers evolution, because siblings vary in their traits and behaviors. You and your siblings, after all, may be wildly different. That's because when two parents' genes join together to make a baby, they do so in unique ways for each offspring. On top of that, when a youngster is forming, genetic mutations pop up—these can be beneficial (like better camouflage for a given environment, maybe), neutral, or detrimental (like a missing limb). This kind of uncertainty leads to organisms of the same species that vary. And because animals produce more offspring than can survive, those with the characteristics that make them best suited, or "fitted" in the "survival of the fittest" sense (not necessarily fit as in jacked), to their environment survive to pass down their genes to the next generation. Over the generations, individuals that fit best in their niche—be it ambling around eating leaves like caterpillars, or wasps doing their best to ruin the lives of said caterpillars—tend to leave more offspring.

So cruelty begets cruelty:* In order to ensure her kids make it in a world full of predators, the jewel wasp has evolved a strategy of extreme brutality. She doesn't have anything against cockroaches, mind you, but she *does* have an overriding instinct to help her kids survive. Though it's not like a jewel wasp showed up one day and started brainwashing cockroaches—the technique evolved slowly, step by step. The wasp might have started out doing what a lot of wasps do, killing the cockroach with a sting and providing the corpse as food for its young. But that would have left the developing larva exposed to predators. So maybe at some point in its evolution, the jewel wasp started paralyzing roaches instead of killing them. That would have allowed the predator to drag the victim into a burrow for the larva to eat in peace. But because the mother wasp is so much smaller than her victims, she would have had to choose small roaches, which would have meant less food for her young. Eventually came the brilliant solution, evolved step by step thanks to random mutations: The jewel wasp invented brain surgery to convince even the biggest of cockroaches to abandon any instinct for self-preservation and walk themselves to their doom.

And the jewel wasp is not alone in its persecutions. All manner of other wasps not only brainwash their victims into serving as obedient meals for their kids, but as an extra indignity will turn the dupes into unwitting babysitters.

---

*To put things another way: The tendency for nature to do very bad things to creatures' bodies is in part what drives the evolution of species.

## If You Be My Bodyguard, I Can Be Your Long-Lost Manipulative Pal

Few creatures enjoy PR as immaculate as the ladybug's. It's an insect, yet it's universally adored because for one, it eats aphids out of gardens. And two, it's one of the few polka-dotted creatures on Earth. Which makes the exploits of the *Dinocampus coccinellae* wasp that much harder to endorse. It stings the ladybug, which then gives birth—and not through the traditional avenues—to the parasite's enormous larva, which is nearly the size of its own spotted abdomen.

*Dinocampus,* like the jewel wasp, doesn't bother with doting motherhood. She drops off her young and flies away. But unlike the jewel wasp, the *Dinocampus* mother has a partner: a virus that she injects into an adult ladybug's abdomen, along with a single egg. That egg hatches into a larva, which begins feeding on the beetle's insides. All the while the ladybug goes about its business—no brain surgery, no imprisonment, just normal ladybug stuff as the parasite within grows. And grows and grows, to incredible proportions. When the orange-yellow blob is finally ready to come out, it forces its way through the ladybug's abdominal segments and into the world, crawling between the legs of its gracious host.

The virus, meanwhile, has gone to work. When the wasp injected her payload, the egg hatched and the virus replicated within the resulting larva, then moved into the ladybug's body—but not its head. Only when the larva is preparing to erupt from the beetle does the virus pour into the brain and replicate once more, leading to a massive immune response that short-circuits the mind. Why the immune system chooses this point to kick in, and not when the giant larva is growing in the abdomen, isn't clear. It may be that the young wasp is secreting something that suppresses an immune response, so once it leaves, the system snaps back on. But what's apparent is that the severe neurological disorder that begins to shake the brain will paralyze the ladybug, just as the larva is settling down between its legs. (This *Dinocampus coccinellae* paralysis virus, as it's formally known, is perhaps not by accident related to the poliovirus, which invades the human brain and spinal cord, leading to paralysis.)

Because this is a brutal world loaded with predators, as the larva spins its cocoon, the virus forces the still-somehow-alive ladybug to do something silly: Instead of bolting, it stands atop the cocoon and begins twitching. Touch it and it twitches all the more. For a week the bodyguard cradles the fiend that erupted out of its body, never once feeding for itself. Dutifully it twitches, kicking like a polka-dotted donkey at ants and other predators that close in. And it's good at its job—one experiment showed that while predators will snatch away every single wasp cocoon left unprotected, and about 85 percent of cocoons topped with dead ladybugs, they get only about 33 percent of the cocoons under the protection of their convulsing guardians. About two-thirds of the wasps survive. The mother wasp has conquered the cruel world with some cruelty of her own.

So the virus transforms the ladybug into a bodyguard, though it's not helping the wasp as a hobby. The guardianship ensures the pupa develops into an adult wasp, free from the harassments of predators. And because the virus can live nowhere else, it depends on the well-being of its wasp host in order to survive. It's a touching partnership that happens to ruin the life of a ladybug. Still, somehow, someway, a quarter of brutalized ladybugs survive the giant larva erupting from their bodies and the near total shutdown of their central nervous systems.

It's a rare happy ending in a world that's reluctant to dole out sympathy and instead encourages a brutality that Darwin and Wallace realized drove the evolution of species. But decades before them, the most famous naturalist of his day, an exuberant German named Alexander von Humboldt, was among the first Europeans to question the prevailing paradisal view of nature. During his own South American travels, Humboldt witnessed the unrelenting suffering of nature's creatures. For instance, the world's largest rodent, the 150-pound capybara (it looks like a scaled-up rat that got punched in the nose), is a semiaquatic creature that spends much of its time wading through rivers with its webbed feet. Humboldt watched the poor things pinball between the terrors of the terrestrial and aquatic worlds, fleeing crocodiles in the water and ending up in the jaws of jaguars on land, and vice versa.

Such quarrels between predator and prey would set off a cacophony in the forest—species upon species like monkeys and birds howling in chain reaction to the disturbance. The locals figured the animals were "enjoying the beautiful moonlight." Your average idealistic European at the time might have figured God had created the chorus for the enjoyment of nearby humans. But Humboldt knew better. "To me," he wrote, "the scene appeared to originate merely by chance, developing into a long-extended and ever-amplifying battle of the animals." Humboldt was toying with what would become the foundation of Darwin's theory of evolution by natural selection: Life sucks, and then you die, usually in a pretty terrible way. Because the fact of the matter is, living organisms are fundamentally fragile things. After all, capybaras are made of flesh, and even crocodiles with their reptilian armor have their weaknesses. Jaguars, as it happens, have been known to sneak up on them and clamp onto the back of their necks, immobilizing the ostensibly impenetrable predators.

This suffering is trivial, however, compared to the adversity of the tobacco hornworm caterpillar, among the most fragile of fundamentally fragile things. It endures an even more torturous fate as an unwitting bodyguard, thanks to a wasp that injects it not only with a virus and an egg like *Dinocampus* does, but with up to two hundred eggs that will hatch into larvae and erupt out of its flesh simultaneously. But that nightmarish trauma isn't what the caterpillar has to really worry about.

## The Very Not Hungry Caterpillar

On the outside, the infected caterpillar is still a charmer, all green with delicate white stripes and a peculiar little horn coming off its bum. Hence the name hornworm. It inches around in the sunshine doing normal caterpillar things, which generally amounts to eating leaves. On the inside, though, it's chaos. Hundreds of wasp larvae squirm about, soaking up fluids—not chewing on organs, mind you, but generally guzzling. Their mother dropped them off a week earlier, and by that I mean she stung the caterpillar and injected her eggs into its body cavity. She also pumped in a venom loaded with her own brand of virus found only in insects, what's known as a polydnavirus (pronounced puh-LID-na-virus, by the way). That'd be the expeditionary force. The virus's job is to beat back the immune system so the caterpillar's body doesn't destroy the eggs before they hatch. And the virus is good at its job: It kills hemocytes, cells in the caterpillar similar to our own white blood cells. The immune system thus shuts down and the eggs hatch unimpeded into larvae, growing and growing until they stuff the caterpillar's body cavity.

But there comes a time in every *Cotesia congregata* larva's life when it must leave the tobacco hornworm. So all at once, the larvae start gnawing through the caterpillar's fat on their way to the skin, which they scrape to open up their own exits all over the body. Soon enough, the larvae are squirming out of the caterpillar, which doesn't appear to mind at all—the youngsters, it seems, are secreting an anesthetic to numb the holes they make. Indeed, poke at one of these wounds and the caterpillar won't protest, but poke the skin a few millimeters away and it flies into a frenzy like normal (caterpillars being sensitive about these sorts of things). But why doesn't the caterpillar bleed to death from this massive trauma? you might ask. Because the wasp larvae shed their exoskeletons from time to time as they grow. And their last molt is timed with their emergence from the caterpillar, so the discarded exoskeletons plug up the wounds.

Once outside, the larvae stay on their host's body and spin their cocoons, attaching to the skin with a bit of silk. What you end up with is a caterpillar covered in little white ovals. It's sitting still, and even weirder, it's stopped eating. Because a caterpillar's only job in life is to feed and feed and feed to fuel its growth and eventual transformation into a butterfly or, in the case of this tobacco hornworm, a moth.

"This is a very profound change in an animal that's literally an eating machine, that's what caterpillars do," says biologist and *Cotesia* aficionado Shelley Adamo, who keeps a colony of the wasps at Dalhousie University in Nova Scotia. (Well, normally she does. I pestered her to let me visit for almost a year, but her wasp import permit—yes, a thing—had lapsed. She was patient enough to let me call her instead.)

So the caterpillar doesn't whip around and devour the parasites stuck to its body, which seems like an oversight. It's as if it has given up. And in a way it has—on a deep, physiological level.

Let's back up a bit. *Cotesia* has itself a challenge that the jewel wasp doesn't so much have to worry about: its host's immune system. The jewel wasp spends the bulk of its time outside its victim's body, but from its earliest days *Cotesia* has to somehow hide from patrolling immune cells. "The larvae actively do something to prevent the host immune system from recognizing them," says Adamo. What, exactly, isn't certain. "But the eggs don't have that ability, which is why the polydnavirus needs to knock out the immune system temporarily." Yet the larvae and virus can't risk shutting down the immune system—if the caterpillar can't fend off other parasites, it'll perish, and the baby wasps will go along with it. So they're striking a precarious balance here.

But then the larvae drill out of their host, setting off something known as a cytokine storm. Cytokines are molecules that allow immune cells to talk to one another, and they go into a frenzy once the parasites exit the caterpillar. In fact, if for whatever reason a larva doesn't make it out with the rest of its brothers and sisters, the immune system will roar to life and mob and destroy the holdout immediately. The caterpillar's defenses, it seems, are back online—only it's too late.

So the caterpillar is soaking in cytokine and also in irony, for this immune response will be its undoing. The problem with cytokine is that it also tends to dampen appetite. This loss of appetite due to sickness or parasites, known as illness-induced anorexia, is not unusual among animals. But the caterpillar's body gets a bit carried away with it, to the point where it will never feed again and thus starve to death, only after the wasps have transformed into adults and flown away. Which is what the pupating wasps "wanted" all along: A caterpillar without an appetite is a caterpillar that won't devour the cocoons stuck to its body.

"But they also don't want their host eating anyway," Adamo says. "Because if the host feeds, even if it doesn't feed on them—it just feeds on leaves—some predators can home in on plant volatiles." These are noxious compounds that some plants release to ward off foes and attract predators to take care of whatever is gnawing on their leaves. Think of it like a distress call. Indeed, mother *Cotesia* found her once-hungry host in the first place by sniffing out these volatiles.

Compounding the caterpillar's problems is something called octopamine. This is a neurotransmitter that neurons use to send messages back and forth in an insect's brain. And octopamine, too, goes haywire as the larvae are emerging, spiking to five times normal levels. Which is appropriate given it's a stress hormone, and the caterpillar's body is just now realizing it's under a good deal of stress. Unfortunately for the caterpillar, though, octopamine further suppresses appetite.

"You can understand if an animal is fleeing," Adamo says. "Some horrible thing is happening to it, and often when animals are under that kind of situation they don't eat because they're too busy defending themselves. So octopamine probably for that reason has an antifeeding effect." Between the flood of octopamine and the cytokine storm, the caterpillar is nutritionally screwed. So it sits there as an obedient bodyguard, batting away ants and other insects that try to make off with the cocoons.

Think again about the chain of events that unfolds here. The *Cotesia* mother first injects her venom and the virus. The latter beats back the immune system, and the former seems to stop the caterpillar from ever metamorphosing into a moth. Indeed, inject caterpillars with only the venom and virus without any eggs, and the things will swell like balloons. "They become *enormous* caterpillars, the biggest caterpillars you've ever seen," Adamo says. "Because they just keep growing and eating until they finally die." This arrested development makes solid sense. After all, the mother wasp needs to give her kids enough time to develop and exploit the caterpillar as a bodyguard. If the caterpillar manages to transform into a moth, they all die. And if the larvae can't manipulate the bodyguard into defending them, they all die. (Leave cocoons alone on a leaf and every single one will perish. Again, life is a bummer, hence all the effort on the wasp's part.)

What's so magnificent about their exploits is that they have to set them in motion *before* they've left the caterpillar—remember that the cocoons remain attached by only silk, so they're no longer tapped into the host's system. The thinking here used to be that a few larvae would stay behind to brainwash the caterpillar, but not so. Even when all escape, still the host transforms into a bodyguard—that and the caterpillar's immune system comes back online to kill any holdouts anyway. So however it is they're glitching the caterpillar's octopamine and cytokine levels, the larvae are doing it while they're still inside the victim. If they've done something even slightly wrong and the events don't cascade like normal, they all die.

The moral of the story being: You have no right complaining the next time a wasp delivers you a little sting. Among the many zombifiers we'll meet on this journey, the cruelty that wasps inflict is unrivaled—they've got instinctual schemes and stingers loaded with venom and eggs. While scientists are still trying to work out the chemicals and neuroscience at play here, what we know for certain is that such intricate zombifications are the product of the simple processes of natural selection, of species struggling to survive a brutal world. Those that can tough it out will survive to pass down their genes. So generation after generation, parasites build up manipulations so elaborate, it's hard to believe a god didn't dream them up. Such is the awesome power of cruelty.

Now, I can't begin to imagine what it feels like to have a parasite consume me from the inside out. It's something the human race will never know, like why exactly it isn't okay to wear white after Labor Day. But what's even more unfathomable are the zombifiers that not only consume the insides of their hosts, but grow as a network through muscles and release chemicals in the brain to precisely steer their dupes around. Oh, and to do all this without a brain of their own. You see, the zombifiers out there don't just drag their victims into a den or freeze them in place on a leaf—they turn their victims into their own personal vehicles.

# 2

# Nothing Brings the World Together Like Unsolicited Mind Control

***If you thought you had nothing in common with a fungus that invades an ant's body and takes control of its mind, you'd be incorrect.***

Sporting bright blue eyes and matching T-shirt, David Hughes leans back in his office in a standard-issue professorship chair, as Penn State students in a plaza behind him shuffle toward classes. Between us on his desk—on either side of a paper cup of black coffee—are two trays of erstwhile ants stuck through with pins. Some cling to leaves, others curl up around sticks, frozen in their death postures like the now-fossilized humans who couldn't escape Pompeii. All, though, have strange structures erupting out of their corpses.

I'm going to be honest with you: The zombie ants are smaller than I expected. I understand that they're ants, and therefore small by default. But maybe ever since I informally met them in my kitchen three years prior I'd kept building them up in my head. Now they just seem so dainty and vulnerable, the fungal stalks coming out of the back of their heads looking more like threads than, well, stalks. Zombies are supposed to be more . . . robust, right? More indestructible?

As I'm grappling with the reality of ant size, Hughes pivots his monitor to show me a microscope photo of an infected ant's muscle. More specifically, an ultrathin slice of an ant muscle, so the blobs we're seeing are cross sections of fibers. In between these blobs, though, are tinier blobs—fungal strings that have grown through the muscle fibers, prying them apart. Photograph slice after slice of an ant muscle like this, use AI to detect the bits of fungus and paint them green, and stack the photographs once more to make a 3-D model, and you can start to grasp the destruction the fiend has wrought. What Hughes has imaged is a muscle overwhelmed by fungus, insidious strands of green growing like grass between the fibers.

This isn't like the wasps. This isn't just deploying some sort of chemical that takes control of the host's brain. This is something more complicated: A fungus called *Ophiocordyceps*—to be referred to henceforth as *Ophio* for the sake of brevity and to cut down on misspellings—infiltrates and hijacks the ant's muscles, but doesn't touch its brain. "It's basically just punching holes in the muscle," Hughes says, pointing to said muscle. "So this is really strong atrophy, the same kind of stuff which would happen if you broke your spine."

The fungus severs the nerves in the muscle, cutting off communication with the brain. Essentially, crippling the nervous system. And that wouldn't seem to make a lick of sense, because up until the end, the fungus isn't paralyzing the ant, but instead assuming precise control over its faculties.

"If I had to guess, and this is completely speculative," Hughes mentions casually, "I think the fungus is forming a nervous system."

## The Zombie Ant:
## Chronicle of a Death Foretold

In the forests of the Americas and Southeast Asia, you'll find an estimated five hundred species of *Ophio*, each targeting only one species of ant. If the fungus gets into the wrong body, it's a literal dead end—the parasite will perish. It can grow in there all it likes, but it won't be able to manipulate the ant's brain: The key won't fit the lock.

This hyperspecialization might appear a weakness—after all, why settle for one species when you could exploit any number of others in the forest? But do keep in mind we're talking about assuming control over the incredible complexity of the brain, which even in a little ant is wildly intricate. Intricacy, then, probably demands specialization, a particular *Ophio* species evolving in tandem with a particular ant species over the millennia.

This tale begins with the humble *Ophio* fungal spore. Once it hits the ground, it sprouts a secondary spore that grows vertically, tipped with a bit that catches on an ant's exoskeleton. "And these have backward-facing spines in some species," Hughes adds. "So they just attach"—[*emphatic clap*]—"like a limpet mine. And then they literally blow a hole and eat through the cuticle at the same time." That's thanks to enzymes that break down the ant's armor, plus the spore builds up pressure to equal that in the tire of a 747.

*Kaboom*. The fungus is in.

And what a wonderland it has found, for the ant is not just its vehicle, but an energy-rich battery. An insect is not built like us—it has an open circulatory system, so things are more or less sloshing around in the absence of a network of arteries and veins. It's not hard, then, for the fungus to get about. As it multiplies, it soaks up more and more of its host's nutrients. Exploratory bits of *Ophio*, known as hyphae, branch out to find ever more food, growing as a network throughout the body. Fungal cells infiltrate the muscles, breaking the fibers apart. All the while, the fungal colony is talking, forming so-called anastomosis tubes—think of them like pneumatic tubes, only without the vacuum.

"This fungus has joined together in a group and they're communicating and they're exchanging things, that's what anastomosis is," Hughes says. "The question is what they're exchanging and what they're doing. We don't know." It may well be food, the exploratory bits ferrying nutrients throughout the network.

Regardless of what the network is doing, it continues to multiply, spreading farther and farther, eventually reaching the brain, where it . . . stops. The *Ophio* fungus, for all its powers of mind control, never invades the ant's brain. Instead, it grows around the thing as a sheath. After all, the fungus wouldn't want to risk crippling its vehicle—it lacks the surgical precision of the jewel wasp.

After three weeks of growth, the fungus will make up perhaps half the insect's weight and is now ready to flip the switch. Yet all the while, the ant has acted normally—no stumbling, no aggression—nothing that would alert the colony to an intruder. The fungus within has been able to diversify into different tasks. Some bits mine nutrients, some attack muscles, and some surround the brain, ready to release a chemical bomb.

When that bomb drops, the *Ophio* fungus manages to convince an ant to do the unthinkable: not only leave its beloved colony, but sabotage it. Away from the watchful eyes of its comrades, the ant now begins staggering. It convulses its way up a tree trunk, tripping along a branch and onto a leaf, sinking its mandibles into the vein. The fungus having taken control of the mouth muscles, the ant assumes a death grip. Six hours later, the victim perishes, and the fungus consumes what little remains of its insides.

Critically, the fungus has steered the ant ten inches off the ground, where the temperature and humidity are best for the astonishing growth it's about to undertake. The *Ophio* has to work quickly, for the rain forest is teeming with all manner of other fungi (perhaps forty thousand spores of various other species on a single leaf) and bacteria that would happily consume what remains of the ant. But the parasite is prepared: Cross-section an ant at this point and you'll find that among the white hyphae that have replaced the ant's insides is a band of orange. This is packed with carbohydrates, and probably feeds the maniacal growth of the fungal stalk that erupts out the back of the ant's head. Hyphae also pour out of the ant's mouth, further securing the mandibles to the leaf, and proliferate as a sort of fuzz on the ant's cuticle, protecting the prize from invaders. Everything thus in order, the *Ophio* stalk matures and begins releasing spores, which descend onto the hapless ants marching below.

## Your Long-Lost Uncle, the Fungus

If you find this all to be unbelievable, good. I once felt the same, so we have something in common. It's your skeptical human mind questioning things, even though you have no reason to in this case because I would never lie to you. But the idea of what is essentially a hybrid creature—part fungus and part ant, two unrelated species coming together to create something that's difficult to define—is hard to come to terms with. We modern humans know that organisms are split into kingdoms—plants and animals and fungi have nothing to do with each other. Our bodies have nothing in common with *Ophio* or ants or manipulative wasps.

Except that's not really true.

Let's take a lesson from the Middle Ages. Between the thirteenth and seventeenth centuries, Europeans had this funny idea in their heads. For millennia, they'd made their clothes out of sheep fuzz—that is, wool. But then something newfangled arrived when eastern trade routes opened up: cotton. Europeans, even the educated ones, didn't know what a cotton plant looked like. They'd just heard that in the East, wool grows on trees. (The ancient Greeks were plenty aware of cotton, thanks to their forays into the East. But that knowledge was largely lost as Europe grew more insular.)

And so began perhaps the most hilarious large-scale misunderstanding in history: the so-called vegetable lamb of Tartary (a region that is now central Asia and Russia). Word spread that cotton came from lambs that grew on trees in one of two ways. The first looked like a true cotton plant, only instead of cotton erupting from the buds, tiny lambs would grow, complete with legs and eyes and ears and all that. The second version of the tale was even more outlandish: A single lamb would grow on a stalk like a sunflower, connected at its belly as if by umbilical cord, its limbs dangling. Leaves grew around the base, so whenever the lamb got hungry, it would bend down on its stalk and feed. Tragically, though, once it had consumed all the vegetation, the critter would starve to death.

Medieval Europeans had invented a hybrid creature, part animal and part plant. I tell this tale not to make fun of people who had no internet access to fact-check this stuff—I tell you this because the idea of a hybrid creature isn't so outlandish. No, lambs don't grow out of the ground. But two organisms that couldn't be more different or distantly related, the ant and the *Ophio* fungus, have fused together to form what you may as well consider a new species, one even more improbable than a vegetable lamb. We're talking about a creature that is entirely unlike a fungus and entirely unlike an ant, yet is both.

We might ask, then: What *is* a zombie ant? Is she still an ant if she's actually more fungus by weight? Is she still an ant when *Ophio* convinces her, one of the most fiercely loyal creatures on Earth, to betray her comrades? After all, she works like hell her whole life, until a spore happens to land on her body. "And then twenty-one days later you come out of the colony and you die above your sisters and infect them," says Hughes. "So at some point you no longer are an ant. At which point does that change, and how can we recognize it? What are the morphological cues, or what are the genetic cues?"

In the end she looks like an ant, but she's been reprogrammed into a different kind of being: a fungus wrapped in what you and I—and more critically, her sisters (workers are all female)—would recognize as an ant. She dies not as herself, but as a costume for a manipulative parasite. Which concerns me. I like the idea of an ant *qua* ant. I like the idea that a creature I see in nature is actually the creature it's advertised as. But *Ophio*, more than any other zombifier we'll meet, gets full-on *Invasion of the Body Snatchers* with it, manipulating not specific behaviors, like the wasps that get caterpillars angry, but essentially replacing the host's mind with its own, well, *not* mind. It has no such thing, yet it manages to brainwash an ant.

Now let me pile on the existential dread for us. In the grand scheme of things, you're not all that different from the ant or the fungus. No, you don't have six legs, or zero legs for that matter, but all three parties, in addition to every species that's ever lived, descended from a single common ancestor, the first organism to appear on Earth some 3.5 billion years ago. All of us—humans and plants and insects and fungi and, yes, even monstrosities like pugs—share a common language: DNA, the genetic code that programs our bodies and behaviors. When early life began to speciate, and then speciate and speciate some more into the incredible diversity of organisms that now walk and crawl and swim this planet, that genetic code followed along.* It is the

*Speciation is usually the product of isolation. Say a new river cuts through a valley, splitting a population of wild pugs in half. (This is my story, I can tell it however I like.) The two isolated populations will begin to diverge genetically, perhaps adapting to the particular conditions on their respective sides of the river. Give them enough time and the pug clans may

language that binds what you might consider a "higher" life form like a human with "lowly" fungi. We're all one big family, however different we may seem, united under a universal code of life.

It's this shared system of DNA, not to mention proteins and hormones and neurotransmitters, which we subscribe to that makes the exploits of the zombifiers possible. The jewel wasp didn't invent some novel compound to shut down the legs of its victim—it co-opted GABA, a common inhibitory neurotransmitter. And the *Cotesia* youngsters set off a cytokine storm, turning the caterpillar's own body against it. *Ophio*'s tricks, too, have an explanation. Now David Hughes has to find it.

## Hey Sister, Go Sister, Soul Sister, Go Sister: Ain't No Family Drama Like Ant Family Drama

But first things first. I don't want to question *Ophio*'s methods, but wouldn't it have been a whole lot easier to skip hijacking ants' brains and just muscle into the colony itself, where there are plenty of individuals to infect? Sure, if it weren't for ants not taking kindly to such things.

To fully comprehend the majesty of fungal zombification, we have to first comprehend the majesty of ant society. Ruling the colony is a queen. The vast majority of the young she produces are females—the workers who dedicate their lives to the queen and each other. Cooperation drives the whole operation, the ants foraging and sharing resources and fighting other colonies to protect their territory. Thus the colony works as a kind of superorganism,

---

grow so genetically distinct that they can no longer breed with one another to produce fertile offspring. And there you have it, two new species that probably still have trouble breathing.

thousands of individuals banding together like cells to create one body in pursuit of reproduction on a grander scale.

Problem is, each cell is an entry point for parasites with cruel intentions, so our superorganism needs an immune system of some kind. What the ants have landed upon is called social immunity: Each individual acts as an immune cell to detect and eliminate intruders. So if someone is acting oddly—say, staggering around—that's an indication that she's host to something nasty that could destroy the colony if not quarantined. A worker will pick her up, drag her outside the colony, and unceremoniously dump her in a graveyard of other diseased ants.

This system presents a problem for *Ophio*. It has to grow within an ant, yet not alter its host's sociality, a tall order when it's wrenching apart muscle fibers. So somehow the fungus is able to time it such that the ant begins stumbling only once she's on her way out of the colony and up the tree. Not only that, but *Ophio* has to avoid the colony's immune cells literally sniffing it out, for ants use pheromones—or chemical cues—to identify each other and communicate things like the location of food. Considering the fungus can make up half of an ant's body, that's one serious feat.

So in order to get anywhere at all, *Ophio* has evolved to fool the superorganism that is an ant colony, and its solution is body snatching. If it made the ant start acting in any way funny, it'd be evicted. If it made the ant smell weird, it'd be evicted. And if it dispatched the ant within the colony and started growing a stalk, it'd be evicted.

Appreciate all the factors involved here. The fungus has to grow at the right speed, lest it overwhelm its host and drop her dead. It has to mask its own chemical signature, lest sentries sniff it out. It has to avoid destroying the host's organs. It has to avoid growing through the cuticle. And it has to invade the muscles yet keep them operating, perhaps by releasing some kind of neurotransmitter that mimics the ant's own to control the fibers. It's the most literal manifestation in the animal kingdom of a puppet master pulling strings. And like any halfway decent puppet master, the parasite has to stay out of sight, or the whole masquerade comes crashing down.

So we have a good idea why *Ophio* might have evolved into a zombifier. But *how* could it have done so?

Over the millennia, random mutations allowed *Ophio* to conquer the social insects by stepwise exploitations of their defenses, developing an increasingly sophisticated manipulation to drive the infected ants out of the colony and up into trees. I'm going to speculate here, because no one without a time machine knows exactly how this evolved, but perhaps it was that the fungus began by invading ants and killing them wherever, maybe while the victims were still in the colony. Not great for the fungus's development on account of the temperature and humidity in there, and not great for not getting found out. Then certain mutations gave the fungus some measure of control over the ant's behavior, allowing it to guide the host at least out onto a trail to die and release its spores on the path. Then another mutation and the ant was up a tree. And another and she was on the underside of a leaf.

Again, that's speculation, but whatever increased the fungus's odds of replicating got passed down to the next generation and the next over countless millennia. A whole lot of spores ended up going nowhere, perishing in the hostility of nature, but those that did persist were the best "fitted" to their environment because they could brainwash their hosts. Over time, tiny changes added up to produce this astoundingly complex manipulation. So it wasn't as if a fungus showed up one day and started bossing ants around. It was a slow, methodical (yet undirected) march toward mind control that helped the parasite reproduce.

Maybe the key was the muscles. Maybe *Ophio* started out consuming those tissues and eventually figured out how to manipulate them instead, prying the fibers apart, crippling them. "What happens with rigor mortis is the calcium stops flowing in and out of your muscle cells," Hughes says. "And I think what they've done is induce a functional rigor mortis in a living ant." It's the real-life living dead.

This means the fungus doesn't need to dig its way into the brain, but that's not to say *Ophio* isn't in some way affecting the ant's mind. Indeed, looking at the fungus's evolutionary tree could give us some insight into what's happening here. Namely, *Ophio* is closely related to a fungus called ergot, which in 1938 a Swiss scientist by the name of Albert Hofmann synthesized into lysergic acid diethylamide.

But you probably know it as LSD.

## The Fungus Among Us

In August 1951, the tiny town of Pont-Saint-Esprit in southern France—a place not known for heavy drug use—had itself a mass freak-out. Folks started getting burning hot or freezing cold, developing severe digestive issues, going into convulsions. And then came the hallucinations. As one man claimed: "I am dead and my head is made of copper and I have snakes in my stomach and they are burning me." Then he tried to throw himself into a river. An old woman launched herself into a wall, snapping three ribs. Others tossed themselves out of windows. The more cautious among the townspeople hid from their hallucinations under blankets.

The problem was likely a bad batch of bread, which a local baker had made with tainted flour. Authorities fingered the culprit as a fungal parasite of grains—ergot—whose spores blow in the wind and attach to the feathery bits of the flowering grain, then grow as hyphae deeper into the plant. As the fungus matures and replaces ever more of the grain with itself, it produces a sweet secretion called honeydew that's loaded with spores. Insects lap this up and transport it to other grains, thus spreading the fungus across whole fields.

Ergot packs that hallucinatory, toxic kick as a defense against herbivores, which will either learn their lesson not to graze on grains or simply drop dead. Whichever way, the fungus is trying its best not to end up in a stomach, where its life would come to an abrupt and acidic halt. (Ergot infection is therefore both a blessing and a curse for grains, oddly enough. If the ergot doesn't end up killing the plant, it gifts the host with a powerful weapon against grazers.) Ergot, then, evolved to target mammalian nervous systems, growing more powerful over the generations, natural selection having favored the most toxic fungi. Specifically, the fungus is producing alkaloids that overstimulate the central nervous system—hence the convulsions and hallucinations and sprinting into walls.*

---

*Ergot isn't *entirely* bad news for humans. As the sixteenth-century doctor Adam Lonitzer noted, midwives could use it to induce contractions during childbirth. And if you thought *Ophio* is striking a fine balance by manipulating an ant just right, you've never been a sixteenth-century midwife: Doctors grew wary of the practice when they realized it was hard to figure out the correct dosage. Administer the right amount and you'd get a baby—administer too much and you'd get a convulsing mother.

These days, ergot-derived drugs like Migranal can relieve migraines by constricting blood vessels in the brain, while ergot-derived drugs like LSD can relieve reality by constricting the nature of time and space.

Now, we can't begin to know if an *Ophio*-infected ant can hallucinate, much less if those hallucinations would involve snakes burning up their stomachs and turning their heads to copper. But what we certainly know is that as zombified ants are leaving the colony, they're convulsing. And that if you were to open up one's brain and take a tissue sample, you'd find fifty-five times the normal amount of ergotlike alkaloids, which resemble neurotransmitters. So the *Ophio* cells that surround the brain are probably dosing the ant—big-time. But again, the fungus is striking a balance here: It has to drive its host mad, yet not mad enough that the colony raises the alarm. And the bits around the brain releasing the chemicals have to communicate with the rest of the fungal network throughout the ant without sending the whole system into disarray. Remember that not everyone has the same job—some of the fungus soaks up nutrients and other hyphae pull the strings in the muscles. "It makes a lot of sense," Hughes says. "If you're in the back of the bus, you don't need to produce the chemicals that affect the driver. You're just waiting until the driver is manipulated."

What makes *Ophio*'s powers of manipulation all the more mysterious is the fact that not every zombified ant crawls up above the colony's established trail. Some wander far and wide, cast out of the colony like asteroids flung out of a solar system. They stumble on and on, farther and farther from the protection of their comrades. But why would the fungus risk this when it has a perfectly good colony to exploit? The answer, Hughes thinks, might be that because these ant colonies don't last forever—the queen inevitably dying and society collapsing around her—the fungus is dispatching scouts. "Probably part of the strategy is to send some of your individuals out into the forest in a long, lonely walk. They may or may not succeed, but they also may get by happenstance to some small colony which is just beginning." This isn't intentional, of course. It may be that there's some variation in how the fungus affects its hosts, not to mention variation in how those hosts react to the infection.

One particular brand of *Ophio*, which calls the United States home, goes about things even more cleverly. Because it shouldn't make sense that *Ophio* could persist so far north of the tropics, where in the winter fungus-nurturing conditions fade away and ant-bearing leaves shrivel and fall from trees. Yet here in the forests of the US, the zombie ants perish in vast killing fields.

Hughes put me in touch with the woman who discovered this American variety of *Ophio.* Not a scientist, as it turns out, but a homeowner with a particularly restless dog. In 1997, Kim Fleming moved to her great-aunt's eighty-six acres in Abbeville County, South Carolina. There's a nice pond there, and a hardwood forest that makes up about half the land—oak, hickory, elm, and the like. Cows once grazed the other half, but the trees are slowly coming back in those bits.

In 2004, Fleming bought a Canon DSLR—with macro lens—so she could snap pictures on her daily walkabout through the trees with Ruby, her Russian wolfhound. Insects, plants, what have you. And, eventually, her first shot of an imperiled ant, clamped onto a tree. (To reiterate, zombie ants are small—smaller than you'd think—hence the macro lens.) "I had seen photographs of them in magazines, and I thought, well, they only existed in more tropical climates," she tells me in a marvelous drawl. "So I was pretty surprised, and really excited, too. But I just kept posting pictures on my Flickr account."

That eventually got the attention of Hughes, because quite frankly zombie ants don't belong in South Carolina. So he asked if he could pay Fleming a visit. "I took him out walking around the woods, just showing him, hey, they're here, here, here, here.

They're almost everywhere you look. It was that day that he was here, he made a proposal. He said, 'If you will gather some data, take some photographs, we'll do a little collaboration.' And he said, 'I'll give you a coauthor credit on whatever papers we do.'"

So their first paper introduced the world to a peculiarly peculiar kind of zombie ant. Because in South Carolina, where leaves fall from trees in the winter, the zombie ants Fleming documented weren't biting into the veins of leaves and perishing—they were biting onto *twigs* and perishing. By ordering its ant to take hold of a twig, this variety of *Ophio* ensures it maintains its perch no matter the time of year. That's particularly important considering that the lower temperatures in a temperate clime retard the growth of the fungus. Since 2004, Fleming has been monitoring hundreds of ants, and some of these zombies are developing over the course of years. The fungus's growth slows in winter, and it sometimes even ices up and goes into a kind of stasis, yet revives in the spring and continues development, all so it can one day ruin the lives of other ants.

## Attack of the Zombie Unicorn Caterpillar and the Flying Salt Shaker of Death

Read enough about *Ophio*, talk to enough people about *Ophio*, and you'll start hearing that this fungus isn't the only game in town. It has an extended fungal family that manipulates a range of insects, from flies to cicadas. And like any good family, each member has its own way of irritating the living hell out of its victims.

On the plains of Tibet and Nepal, for instance, a strange kind of grass sprouts. It's light brown, about as thick as a matchstick, growing bulbous and then tapered toward the tip like a flame. It can be difficult to spot, but should you so happen to find one, you'll want to pluck it out of the ground. After all, a pound of this stuff could sell for fifty thousand dollars.

When you pull it up, you'll know right away you're dealing not with grass but with a fungus and a subterranean caterpillar most unfortunate. The blade is in fact the stalk of *Ophiocordyceps sinensis*, which has invaded the ghost moth caterpillar's body, consumed its tissues, and grown as a unicorn horn out of its head, poking through the soil and into the light of day. What's odd about this isn't that the caterpillar is burrowing underground—that's allowed and indeed normal for ghost moths—but that the fungus has positioned its host so precisely, driving the thing near the surface and angling the body vertically before killing it and erupting as a stalk. It's a manipulation that, like with ant-attacking *Ophio*, boosts the fungus's chances of finding new hosts. *O. sinensis* doesn't have to worry about getting evicted from a colony, sure, but it does have to worry about getting above the surface, where its spores can spread far wider than if it were trapped underground. It's powerful evidence that *Ophio* is a supremely adaptable fungus, on one end of the spectrum commanding a caterpillar to dig toward the surface and die vertically, and on the other end pulling off the more convoluted march of the zombie ant.

Actually, I take back what I said about you plucking one of the caterpillars out of the dirt. Leave it alone, because it's probably not worth the money. These things—known locally as *yartsa gunbu*, or "summer grass winter worm," on account of the fungus infecting the caterpillar in winter and maturing in the summer—are so valuable that people murder each other over them. Prime territories of fungal grasslands turn into battlefields—in 2007, a turf war erupted between two villages near Tibet, ending with eight people shot dead. In 2013, two died in another battle. (The

Dalai Lama, not typically one for negativity, has called such quarrels a "disgrace to the Tibetan people.") In the Chinese town of Chengdu, burglars once tunneled into a store, relieving it of $1.5 million in zombie unicorn caterpillars. Why that hasn't been made into a movie yet, I haven't a clue.

*Yartsa gunbu*, the story goes, can cure all manner of ills, including pain, asthma, and, most covetously of all, impotence, earning it the moniker of "the Himalayan Viagra." (A Chinese track and field coach once claimed that two of his athletes bested world records at the country's 1993 National Games thanks to a steady diet of the stuff.) Science has yet to prove out those bold claims, but the fungus does indeed appear to have certain medicinal charms. It produces antimicrobial compounds, for instance, to ward off its own attackers as it erupts from the caterpillar's head. *Ophio* species that target ants are doing the same thing when they kill off their hosts, releasing massive amounts of antibiotics to beat back any number of bacteria in the rain forest that would otherwise consume them. And such a defense also turns out to be of use for humans: Cordycepin, a drug originally derived from the insect-hunting fungus *Cordyceps militaris*, can boost the human immune system. Which means that if a *yartsa gunbu* brawl lands you in the hospital, it could be a fungal relative that saves you.

Now, the ground isn't the only place loaded with killer fungi. If you're a common housefly, the sky isn't safe either. A fungus called *Entomophthora muscae* (meaning "insect destroyer of the fly," because redundancy be damned) targets these fliers, growing throughout their bodies like *Ophio* does to ants. But this manipulation is a bit different. Instead of ordering its host to climb up onto a leaf and bite down on the vein, *Entomophthora* steers its fly to the tip of a blade of grass. Here the fly, struck with what's known as summit disease, splays out its legs and fastens its proboscis—a sort of vacuum tube it uses to suck up nutrients—to the surface. Thus secured, the host hoists its abdomen and opens its wings, the ideal posture for the parasite to disperse itself. The yellow fungus grows right through the black exoskeleton, splitting apart the abdomen to create a grotesque sort of tiger striping, raining down spores onto any other flies below.

Like with *Ophio*, this is all happening at a specific hour, in *Entomophthora*'s case in the late afternoon on the fifth or sixth day after infection, so by the time the fungus has killed the fly, it's already well into the evening. That means dew is accumulating, and moisture happens to be great for fungal germination. And if for whatever reason the host and parasite don't get set up until late at night on the fifth day, the fungus will hold off on killing the fly until the next day, when it can ensure it's timing the release of spores just right. Which means the parasite can tell time. It has no eyes of its own, much less a brain, yet it's timing the spore release for when conditions are ideal. So somehow it's tapping into the fly's biological clock, in a sense using its host's eyes to tell night from day in order to set up the execution at the optimal time. How it could do such a thing, though, is a mystery.

As an extra complication, keep in mind here that because this fungus's host is airborne, the parasite has to grapple with what is essentially an extra dimension. *Ophio* guides ants up into a tree, and good on it, but *Entomophthora* has to ensure that after it has consumed a good amount of its host's insides, the insect can still fly through the air and not crash into things. Think if you were on a plane and you ate the pilot's and copilot's guts, then expected to get to your destination.

Not content to spread itself by way of spore showers, *Entomophthora* appears to deploy another manipulation, this one of the sexual variety: Dead, infected female flies grow more attractive to uninfected male flies. This might have something to do with the fact that the female's abdomen swells greatly with fungus, and a bigger abdomen is normally a signal to the male that she might be more carrying more eggs—more potential offspring to ferry his genes into the next generation—making her a more desirable mate. Yet present a male with one infected female and one healthy female of the same size, and he far and away prefers the infected one. So something else is going on here.*

On a hunch, a group of scientists set out to prove that this something is a sex pheromone called (Z)-9-tricosene, which female flies excrete to attract males. The researchers expected infected females to produce buckets of the stuff, but what they found was the opposite: Infected females produced half as much (Z)-9-tricosene as their healthy counterparts, likely because the fungus was ravaging their tissues. It could be, then, that *Entomophthora* is releasing massive amounts of its own chemical that mimics the female fly's sex pheromone—a distantly related fungus reaching across the tree of life to impersonate the scent of an insect. Thus in addition to raining down spores and hoping to infect flies passing below, the fungus more actively reels in sex-crazed males, which not only don't get lucky in the least bit, but get very unlucky in the form of spores stuck to their own abdomens. It will cost them their lives.

When *Entomophthora* infects yellow dung flies (the flies are yellow, not the dung), it does so with even more striking

*What makes this particularly striking is that it contradicts a central law in the animal kingdom: Don't sleep with someone who's racked with parasites. The idea is that you don't want to mix genes with a partner unable to ward off such things, because that would mean your kids might not be able to ward off such things. On top of that, you risk infection yourself.

precision. It will make its host land higher on a plant than uninfected flies typically would, helping ensure the spores have an elevated position from which to fall and infect new hosts. The fungus makes the fly cling to the bottom of a leaf, meaning the descent of the spores is unobstructed like with *Ophio*-infected ants, with the added bonus of providing protection from the rain. (Of particular importance because, unlike with houseflies, *Entomophthora* doesn't order a dung fly to lock its mouthparts onto a leaf, so its position is more precarious.) The fly's wings move down and forward toward the head, in contrast to a housefly's wings simply popping up, thus removing yet another potential obstruction. And, most incredible of all, the fungus steers the dung fly to the side of the plant facing away from the wind, ensuring that instead of the spores blowing back into the plant, they blow off of it and make the journey to flies sitting on neighboring vegetation.

Such precise manipulations demand that *Entomophthora* keep its host in at least a semblance of a working order to position it properly, but another fungus that attacks cicadas isn't so judicious, in the sense that it makes the insects' bodies fall apart while they're still flying. *Massospora cicadina* begins its life cycle as spores in the soil, sticking to young subterranean cicadas as the things crawl to the surface every thirteen or seventeen years, depending on the species, as they are so famously wont to do. From there the spores work their way through the exoskeleton and into the abdomen.

But once *Massospora* gets established in the cicada's body, it doesn't bother with subtlety. It consumes the bug's insides full-tilt, replacing tissues with itself. And I mean *all* of the tissues. That includes the bits holding the segments of the abdomen together, so one by one, from the bum toward the head, the segments pop off, revealing a solid mass of yellow chalkiness that is the fungus. And this is all happening while the cicada is still flying around. Wrote one entomologist scouring the woods around Washington, DC, in 1919: "It was in fact not uncommon to observe an infected individual in which the empty body cavity formed one continuous passage from the last abdominal segment to the head, with two or three of its abdominal segments missing, actively crawling or flying about." Cruelty with a capital "C."

All the while, the hapless cicada is crop-dusting the environment with *Massospora* spores, making it a "flying salt shaker of death," as one observer who unfortunately isn't me coined. It'd be hard to dream up a better way to spread a fungus around. Tumbling through the air, the spores attach to other adult cicadas, which, as you're probably aware, can carpet a forest. (The insects will also transmit the parasite when they attach rump to rump to mate.) But inside these new hosts, the fungus doesn't produce the same kind of spore as it would have had it infected a host coming out of the ground. It manufactures a hardier variety that skips infecting adult cicadas altogether, plummeting to the ground and sitting dormant until, right on time thirteen or seventeen years later, the next generation of salt shakers of death emerge and pick them up.

Different brands of fungi, then, have evolved different strategies for different hosts. *Ophio* has to evade a colony's defenses, so it precisely guides its ant out of view. The ever-valuable *O. sinensis* can afford to manipulate its caterpillar to a lesser—though still impressive—degree. Fungi that exploit flies will mess with their hosts' sex lives. And *Massospora* subscribes to the chaos method: consume a cicada without nuance. Each tactic works in its own way—except, of course, for the host.

And each tactic is a striking reminder that while parasite and host may be separated by eons of evolution, they're still bound by the laws of life set in motion 3.5 billion years ago. A fungus may *seem* alien compared to an insect, but what it's doing to its host is anything but otherworldly. It's futzing with familiar neurotransmitters, or pheromones, or simply rotting away the flesh of its host. Unfortunately for the victims, the collective organisms of Earth aren't that different after all. And the fact that these fungi have all somehow landed on manipulation is telling: Zombification isn't a fluke that's evolved here and there—it's a parasitic proclivity. The strange story of *Ophio* that David Hughes first told me while I paced in my slanted kitchen? That was just the beginning of the scourge.

# 3

# When Life Gets Complicated, Life Gets Zombified

*This is the story of how the lowly worms turned into body snatchers and taught us all a thing or two about the complexity of life.*

So I'm beginning to think the life of an ant isn't easy. Through all its existence it toils to feed its family, only to have a fungus invade its body and steer it to its doom. And it's not just fungi that an ant has to worry about, for an ant is a victim of its own success. Its kind are among the most successful creatures on the planet—conquering virtually every environment and numbering in the trillions worldwide—and that's a serious problem for them, because it means a whole lot of parasites have taken notice.

Speaking of—right on time the ant arrives. She's made her way out of the colony, along a trail, into a field, and up a blade of grass. Here she sinks her mandibles in and waits alone. The sun goes down. The creatures of twilight emerge—mosquitoes buzz about and mice scamper by, much to the interest of the local owls. Still the ant sits tight at the tip of the blade. Hour after hour, waiting, waiting, waiting until the sun comes up.

As temperatures climb, she finally snaps out of it, releasing her grip and crawling back down to rejoin her sisters in the nest. She works as normal through the day, but sure enough as temperatures fall she again abandons the colony, returns to the field, and climbs a blade of grass. Night falls, the nocturnal world bustles, the ant sits tight. The sun comes up and she retreats to the colony and goes back to work. Day after day, she commutes back and forth, until at dusk one day a sheep moseys into the field and hoovers up the ant's blade of grass, unfortunately for both parties involved.

Because growing in the ant's body isn't a killer fungus, but a killer worm: the lancet fluke. Snails expel these in their slime trails as tiny balls, which ants are rather fond of. Shortly after the ant eats a mass of worms, the parasites make their way through the gut into the body cavity and work toward the head. The first one that gets there lodges in the part of the brain that controls the mandibles—the rest of the worms fall back into the body cavity and wait. The driver, or "brainworm," steers the host out of the safety of the colony and into the gut of our ruminant, the only place where its kind can mature. The worm will then lodge in the sheep's liver and trigger significant digestive . . . distress, releasing its eggs for the animal to dump out, oftentimes violently. A snail gobbles up the eggs, which develop into baby worms, which the snail expels once more as tiny balls. These the ants will pick up and take back to the colony to feed their sisters, thus perpetuating the cycle.

Walk through the field of zombies, on down to the bank of a river, and you might find a different species of ant with a different brainworm in its head. In fact, you can spot it from afar, because the worm has severely swollen the ant's abdomen. And this insect is acting even more bizarrely, walking in slow circles out in the open on top of a boulder, hour after hour. Approach the ant and shade it like you're a predator hovering above, and it continues to turn and turn, as if in orbit around a tiny invisible star, not flinching one bit. That's because this brainworm doesn't need to get its plump, lethargic ant into the belly of a grazing mammal, but into a keen-eyed bird.*

---

*The worm may be corrupting the ant's ability to sniff out the pheromones that trace the colony's trails. Without a line on anything, the ant spins helplessly. Indeed, uninfected ants that lose scent trails will stop and circle like this until they can reorient themselves.

Where things get really interesting is when a group of ants lose track of a pheromone trail and fall into a so-called death spiral, one after the other

A worm is a funny thing, or "worm" is a funny word, at least. It can apply to any number of distantly related species, from earthworms to tapeworms. But parasitic worms like the flukes make for particularly fascinating zombifiers because they've got a problem: As they mature, they have to get into different species at different stages of their lives. Some must hop between two creatures, some three, some four. That means steering one host into the stomach of another, where the worm can begin the next stage of its development. So while worms may not have fancy stingers like the wasps, and they can't outright consume their hosts like the fungi, they've mastered the art of transforming other creatures into their own personal vehicles.

Manipulative worms are a supreme manifestation of a historically contentious principle in biology: complexity. Life ought to be as simple as possible, right? You'd figure a parasitic worm would have an easier go of it if it bided its time in a single host without having to expend the energy and resources steering the thing into the stomach of another animal. And think about the complexity in your own body. Your brain is the most complicated structure in the animal kingdom, 100 billion neurons working in concert to drive your convoluted behavior. Since the early days of modern biology, creationists have used this kind of complexity to argue that nothing other than a higher power could create such

---

following one after the other, looping until they all die of exhaustion. The most epic account of the phenomenon comes from legendary naturalist William Beebe, who in 1941 described a spiral of *Eciton* (aka army) ants six columns wide and 1,200 feet in circumference. "Through sun and cloud," he wrote, "day and night, hour after hour there was found no Eciton with individual initiative enough to turn aside an ant's breadth from the circle which he [*sic*—worker ants are females, remember] had traversed perhaps fifteen times: the masters of the jungle had become their own mental prey."

a thing. But it's brains and manipulative worms that prove the opposite: Life is beautifully complicated because life made it so.

## Disco Snails and the Ants That Would Be Berries

You know a human zombie when you see one. The slack jaw, the stiffened arms cocked forward, a missing ear perhaps—typical symptoms of a by-this-point-cliché pathogen ravaging a body. But none of that makes the zombie any better at transmitting the virus to new hosts. What makes that possible is the behavioral modification: the aggression and yearning for human flesh.

Wasps and fungi, then, create traditional zombies, in the sense that they don't give a damn if they visibly wreck their hosts, as long as the vehicles still work fine.* But life for the parasitic worms is more convoluted. They have to not only steer their vehicles, but in many cases give them a flashy new paint job. It's about conspicuity, about ensuring the right predators take notice and the wrong predators stay away. A certain worm has to end up in a certain stomach, after all. Any other gut means annihilation.

No worm goes about this more flamboyantly than *Leucochloridium paradoxum*, in the sense that it turns a snail's face into a dance floor. The party kicks off not unlike any other halfway decent party, when a snail eats bird droppings loaded with worm eggs. An egg hatches and develops into a sac, known as a sporocyst, and lodges in the liver, where it soaks up nutrients through its skin. Eventually it begins branching like roots through the meat of the body, working into the head and landing in the

---

**Ophio*, though, has to at least make sure it doesn't start growing through the ant's cuticle, lest the colony take notice.

eyestalks—rudimentary peepers the snail uses to sense pretty much just light and dark. Here the parasite forms brood sacs packed with larvae, which you can see through the translucent skin of the stalks. It's quite pretty, really, striped green and white and brown with speckles of black. What it looks like to the snail, though, is probably less than pleasant.

And then the show starts. As the snail crawls around, the swollen eyes pulsate, the pattern contracting toward the tip, smushing in on itself, and relaxing again. The colors shift and expand and transform, back and forth in a perverse dance the snail can't stop, for it can no longer retract the eyestalks. Which is particularly worrisome because its eyes now look like two squirming caterpillars. And birds—which the worm uses to complete its life cycle—are fond of such things. Also not ideal for the snail is the fact that on top of hijacking its eyes, the parasite hijacks its behavior, driving it onto a leaf out in the sun—no place for a delicate terrestrial mollusk. Here it moves erratically, at times in circles, a tragic and soon-to-be-blinded vehicle for the worm to get into its next host. (Full disclosure: Scientists are confident this is what the parasite is up to, given how dramatically different infected snails look and act, and how easy it would be for birds to take notice, and the fact that the worms *must* end up in a bird to survive. But thus far, no one has actually seen a bird pick the worm larvae out of a snail's eyestalks, a circumstance that is, according to one pair of researchers, "quite embarrassing.")

An even more improbable parasite is one that, unlike *Ophio*, transforms an ant's body without getting caught in the colony. *Cephalotes*, which makes its home in the hollow tree branches of South America's rain forests, is already a bizarre ant without the help of a manipulative worm. Its soldiers wield enormous flattened heads, and when under attack by other ants, they gather at the entrance and plug up the hole with their noggins, like human warriors forming a wall of shields. Should a lone *Cephalotes* find herself under threat while outside the colony, she'll retreat by leaping out of the tree, splaying her legs to steer herself back to the trunk. So the ants are skydivers, and good ones at that—some 95 percent are able to reconnect with the trunk and scurry back up to the colony.

But *Cephalotes* doesn't have to worry just about other ant species. Like snails, a *Cephalotes* can't help but eat bird droppings, and sometimes those droppings harbor the eggs of a treacherous worm, *Myrmeconema neotropicum*. The ant takes the stuff back to the nest and feeds it to the colony's larvae. Within a larva the eggs hatch, and the resulting worms will breed and survive the young ant's transformation into a very different-looking adult: Instead of coming out black like normal, an infected ant is born with an abdomen that's a striking red and, given its already bulbous shape, looks startlingly similar to a berry. Like, *indistinguishable* from a berry if the ant is resting in a cluster of the fruits. (You might think this would be a nearly literal red flag for the colony, but other ants don't notice the invader. That may be because ants aren't so much visual animals as they are sniffers.) The insect acts eccentrically, too, traipsing around with her bum stuck up in the air, while the female worms inside her brood their eggs. The idea being that if a fruit-loving bird catches sight of the ant, it might make a move. And indeed, the bits holding an infected ant's exoskeleton together will weaken, so the abdomen pops off in a beak.

Oddly, though, *Myrmeconema neotropicum* doesn't need to get into a bird to complete its development. It's already bred right there in the ant. So this manipulative strategy is probably about distribution. First of all, the worm has to somehow get its young out of the ant and into a new larva. What better way to do that than to have a bird package the eggs in a turd? And by hitching a ride in a bird gut, the worm better spreads around the rain forest. Natural selection would have favored this kind of complexity, after all. Worms that went through the trouble of turning their ants into berries to get them into birds were more likely to leave more babies, and therefore more likely to pass down their genes that coded for such shenanigans.

So be it disco or turning ants into berries, the lowly worms have proven themselves more than capable of manipulating their hosts in a variety of complex ways for a variety of complex ends. And no worms among them are more accomplished than the acanthocephalans—the worms that started it all.

## That Time Crustaceans Clamped onto a Guy's Leg Hairs and Began a Scientific Movement

August 1969. Hippies raged on a farm in New York—a mere four hundred thousand of them—for something called Woodstock. Charles Manson and his cult raged as well, though in a different way, leaving seven people murdered in Los Angeles. But up in the relative serenity of Alberta, Canada, two scientists—William Bethel and John Holmes—were wading through lakes in search of parasitic worms that infect crustaceans known as amphipods, critters not dissimilar to shrimp, sans the broad tail. Only problem was, they were shy little animals that skittered about floating vegetation, diving and vanishing into the depths when disturbed.

But the men were finding that a few among the amphipods exercised no such caution. When the scientists disturbed vegetation, instead of diving, the crustaceans elected to cling to the stuff with their claws, called gnathopods. The pair would pull clumps clear out of the water and give them a good shake, yet still the amphipods would hang on. The critters were clinging so tightly, in fact, that if Bethel or Holmes tried to liberate an amphipod from the vegetation, they'd end up ripping off the gnathopods, which faithfully held on as their owner entered the custody of science. If the crustaceans couldn't manage to cling to the disturbed vegetation, they'd grow even more crazed, flinging themselves along the surface of the water, grasping desperately—and they wouldn't stop the madness until they found something to hold. Indeed, as the pair waded through the shallows, amphipods would gravitate to their wake and latch onto their leg hairs.

Standing there in a Canadian lake in the summer of '69, Bethel and Holmes were undertaking the first rigorous study of a parasite that manipulates its host's behavior—in this case a worm called an acanthocephalan (meaning "thorn head" for the spines it uses to drill into its host's guts) that drives amphipods to madness. The pair took infected and healthy crustaceans back to the lab and put them in aquariums with some mud and sticks and vegetation, supplementing their diet with lettuce and brewer's yeast. Above the waterline the men positioned lamps, which created light, dark, and twilight zones in the tank. Whereas uninfected amphipods stuck to the darkness, the wormy amphipods went gaga for the bright zone. And when Bethel and Holmes poked at the infected crustaceans, the things would lose their minds, skimming along the surface trying to cling to anything they could, while normal amphipods would jet to the bottom of the tank, sometimes burrowing into the mud. The scientists had confirmed their suspicion: The acanthocephalan forces the amphipod into brighter waters where it's more conspicuous and toward the surface where it's even more vulnerable to predators like ducks. It's in these birds where—you guessed it—the worm must complete its life cycle.

The thousand or so varieties of acanthocephalan don't limit themselves to terrorizing a single host—these must hop between multiple species. So while the worm Bethel and Holmes chased (or the worm that chased Bethel and Holmes, when you think about it) develops first in an amphipod, then matures in a duck's intestines, where it breeds and releases its eggs for more amphipods to ingest, other acanthocephalans will manipulate different invertebrates as "intermediate" hosts—ones where they can't complete their life cycle—to get into their "definitive" vertebrate hosts, usually a fish or a bird, where they *do* complete their life cycle.

The worms are finicky, though, about what they infest: Like with *Ophio* specializing in specific species of ant, we're talking about precise manipulations here. So an acanthocephalan that pilots amphipods can't pull off the same mischief with a crab. On top of that, the worm has to make its intermediate host more attractive to a specific predator, the definitive host, and ideally no others. Acanthocephalans that need to get into fish, for instance, will steer their amphipods into better-lit waters but not toward the surface, thus avoiding the attention of birds, whose guts would be game over for the worms.

Thus worms with complex life cycles play a dangerous game. They get themselves into a crustacean, fine. They then get that crustacean to a part of the lake or pond where they're more likely to meet their definitive host, also fine. But that doesn't guarantee that the wrong predator won't take notice. So two particularly sneaky acanthocephalans give their hosts an extra nudge.

*Polymorphus minutus*, like so many of its peers, steers its amphipod toward the surface and ideally into the maw of a bird. But another amphipod, known foreshadowingly as the killer shrimp, can interrupt those plans by devouring host and parasite alike. Weirdly, though, infected amphipods are up to 35 percent faster than their worm-free counterparts. And when threatened by killer shrimp, the distance they flee is significantly greater.

The acanthocephalan may therefore be outfitting its amphipod to avoid these predators, while still driving the host toward the surface, where avian predators lurk. This seems to help the amphipod avoid fish predators as well. Infected amphipods retreat to surface vegetation when threatened by sticklebacks, making them—according to one study—far more likely to survive attack than unmanipulated amphipods, which dive instead of ascending when in danger. Thus two complementary pressures could have driven the evolution of the acanthocephalan's manipulation: Steering the amphipod toward the surface kept it out of a dead-end fish host *and* made it more conspicuous for the coveted bird host.*

This is particularly odd for a subtle reason: If anything, parasites are supposed to weaken their hosts by sapping energy, and in some cases growing large enough to weigh their vehicles down. So call *Polymorphus minutus*'s manipulations the *28 Days Later* model of zombification. It was this movie that popularized the

---

*Another worm, *Tylodelphys*, also avoids dead-end predators in a clever way. This parasite needs to get from a fish to a bird. Unfortunately for the fish, that means the parasite takes up residence not in the gut, but in the eye, specifically in the liquid between the lens and the retina. These worms are sizable, and will in fact squirm in front of the retina to partially blind the fish during the day, when birds are on the hunt.

At night, though, the predominant threat to fish isn't a bird, but the longfin eel, in whose belly the worm's life cycle would come to an end. So when the sun sets, the worm retreats from the retina, settling in the bottom of the eye so its host can spot the eels.

notion of the so-called fast zombie. Far from bumbling reanimated corpses, these undead are all the more terrifying because they can sprint after you without their limbs falling off. That's because a virus has supercharged them to its own ends, spreading rapidly to pretty much destroy humanity. *Polymorphus minutus* is up to the same villainy, turbocharging its host to ensure the amphipod gets into the right predator.

This tickles me for two reasons. One, I always thought that making a zombie fast was an interesting twist on the genre, but nevertheless biologically problematic. I mean, if you come down with the flu, you don't exactly get the urge to run a marathon. So thanks to *Polymorphus minutus*, we can vindicate *28 Days Later*—parasites *can* be something other than a literal drag for a host. And two, the supercharged amphipod suggests that nature's zombifiers are pulling off a far bigger constellation of manipulations than we'll probably ever know. Scientists figured out the acanthocephalan steers its host around, and they could have left it at that. But by digging deeper, they discovered a manipulation nestled within a manipulation that would seem to boost the worm's chances of survival. (Being honest, though: Could the amphipod's speed boost be some weird side effect of infection? Yeah, maybe, though keep in mind this isn't how a host that's both weighed down and sucked of energy is supposed to behave.)

The acanthocephalan *Pomphorhynchus laevis* also ups its manipulation in a fascinating way. It needs to get itself into a fish, not a bird, but it doesn't just drive its amphipod into well-lit waters—it makes the thing straight-up attracted to fish. To prove it, researchers divided a tank horizontally with a fine net, and then split the upper half into two compartments with a solid partition. They added infected and uninfected amphipods to the lower half, then a predatory fish to one of the compartments in the upper half. The idea being that the amphipods could move freely under the two compartments, sampling water tainted with the scent of a predator and the untainted control water. And sure enough, infected amphipods gravitated to the fish side, whereas uninfected amphipods stayed the hell away. So by making its host attracted to the predators the amphipod knows full well to avoid, the acanthocephalan is hijacking the very instinct for self-preservation. Imagine as a parallel that the roar of a grizzly bear sounded heavenly to you, not pants-shittingly terrifying. Now imagine how well that would work out for anyone other than the bear.

And the same worm shows that acanthocephalans take their manipulations beyond behavior: Some have the habit of changing the color of their hosts, apparently to make them more conspicuous. That's often a matter of simply tampering with the hue of the exoskeleton as the host develops, but *Pomphorhynchus laevis* goes about things differently. Its amphipod host sports a transparent exoskeleton, which the parasite doesn't bother altering because the worm is already bright orange, so it shines brilliantly through the skin. In yet another clever experiment, scientists set out to prove this makes the host stick out to fish. So they took some uninfected amphipods and painted orange spots on them (a mixture, by the way, of oil paint in the colors of orange and raw sienna, plus some typography correction fluid). While amphipods infected with *Pomphorhynchus laevis* are attracted to bright waters, these painted, uninfected amphipods of course had no such proclivity. Yet they were significantly more likely to fall prey to fish than their ungussied, uninfected counterparts, suggesting color change plays its own role in addition to behavioral change.

Once again, thanks to some clever experimenting, scientists have revealed a manipulation nestled within a manipulation within a manipulation. (An *Inceptulation*? Yes, I'm going to run with that.) *Pomphorhynchus laevis*'s peers get along fine steering their hosts into the depths or into the shallows and leaving it at that. But this worm seems to have evolved to make its host more vulnerable by changing its color and making it attracted to prey. The zombie's all dressed up, with nowhere to go but its doom.

## How to Get into the Business of Ruining Other Creatures' Lives

So, as we've seen, all manner of different worms have developed dizzyingly complex life cycles, changing the behavior and appearance of their hosts to steer them into bellies. And that's a big to-do, adapting not only to survive in multiple species without each immune system obliterating you, but to hack another creature over evolutionary time into serving as your chauffeur. Especially when the wasps and fungi and plenty of other parasites get along fine by sticking to a single host. So why would the acanthocephalans not hang tight in an amphipod, or a bird, for that matter? Why the complexity?

Let's pretend that long ago an acanthocephalan did just hang tight in an amphipod its whole life. It invaded the crustacean and grew at the expense of its host, eventually releasing its eggs with the host's feces. But let's say at some point a villain, in the form of a predatory bird, arrived in the ecosystem. It was picking off amphipods, ending not only their lives but the lives of the acanthocephalans within. Unless, that is, some of the worms were blessed with mutations that kept them from dissolving in the birds' guts.

While pretty much every other animal on Earth at the time was going out of its way to *avoid* ending up in the intestine of another creature, hanging out in a gut would have been a significant advantage for these acanthocephalans. For one, they weren't dissolving like their compatriots without those mutations, which is super. That would have been especially important considering the birds might have been in the ecosystem to stay, making off with more and more amphipods, eventually wiping out the worms that couldn't tough it out in avian guts. And two, a worm with a single host—say, a little amphipod—would have been constrained by the size of its vessel. It could only grow so big without overwhelming the amphipod, so it could pump out a limited number of eggs. But if it got itself into a nice big bird, it got more room to grow and, by that virtue, produce more eggs. That gave it a reproductive advantage over its peers stuck in amphipods. As an added bonus, the acanthocephalans were sitting inside the greatest distribution service a parasite could ask for: As the bird flew around crapping out worm eggs, it was spreading them far and wide—much better than an acanthocephalan stuck in an amphipod could ever muster.

And so a parasite unintentionally complicated its life cycle by adding another host. But really, it pulled off a graceful adaptation in the face of destruction. Over evolutionary time the worm further modified its development; like putting off reaching sexual maturity until it got into a bird. And once it was in league with the predators, it needed to draw their attention. Those individuals that could infiltrate the minds of their amphipods had a better shot at getting into a bird. And some species even went inceptulation, adding extra manipulations that further boosted their reproductive odds. So subtle manipulations added up generation after generation to create an acanthocephalan life cycle that seems too convoluted to be real.

I'll reiterate that this evolutionary story is theoretical. No one knows exactly how the acanthocephalans added a second host and turned into body snatchers. But through phylogenetic work (determining the relatedness of species), scientists have found that acanthocephalans did indeed start out exploiting a single marine arthropod of some sort, then added a predator of that arthropod as a second host. And that, my friends, is how you become a mind controller.

## Evolution: Making Watches (of Sorts) Since 3.5 Billion Years BC

In 1802, an English clergyman by the name of William Paley published a nearly unintelligible argument for intelligent design that went a little something like this. Say you're out for a walk and find a stone. Cool, common enough. But then say you came across a nice mechanical watch on the ground. Open it up and you'd find springs and levers and gears, all working in concert to drive hands that represent the time of day. This intricacy tells us that the watch "must have had, for the cause and author of that construction, an artificer, who understood its mechanism, and designed its use," Paley wrote. "Or shall it, instead of this, all at once turn us round to an opposite conclusion, viz. that no art or skill whatever has been concerned in the business, although all other evidences of art and skill remain as they were, and this last and supreme piece of art be now added to the rest? Can this be maintained without absurdity? Yet this is atheism."

Now, I'm not going to sit here and argue that a watch can spontaneously construct itself, like if you gave a monkey a typewriter and an infinite amount of time it'll eventually produce the works of Shakespeare by accident. Watches have watchmakers. But those watchmakers are human, and to extrapolate that kind of craftsmanship to argue that a higher power *must* be responsible for the complexities of nature is misguided. But hey, let's give Paley a pass—he was writing a half century before Darwin published his theory of evolution by natural selection, which has a good explanation for all of this complexity business.

Here's the funny thing about the natural world: The most "complex" organisms, like mammals, get along just as well as the least "complex," like bacteria. In fact, you could argue that it's not humans who rule the planet, but microbes, which are omnipresent. Regardless, each organism is suited for its particular niche, which demands a certain level of complexity. Bacteria don't need eyes and a brain and a suite of other organs to, say, break down animal carcasses. But we mammals would do well to have eyes with which to see our predators and prey and a complicated nervous system to run it all.

Which is not to say we're special, that we're a triumph of evolution. I like telling people this—because shattering worldviews is a hobby of mine—but we're not the pinnacle, and it's self-righteous to think otherwise. To say humans evolved into perfection would imply there's progress in natural selection, that the goal all along was to start out with the simplest of organisms when life first emerged 3.5 billion years ago, then diversify into more intricate forms until boom, the humans, the Chosen Ones, gained consciousness and took their rightful place as rulers of Earth. Yes, since the earliest days life has grown more complex, critters adding features like senses when natural selection deemed it advantageous to do so. But what natural selection can give, it can also take away. Cave dwellers, like fish and salamanders that live their entire lives in total darkness, over the generations find their eyes fading away. They've grown *less* complex, because if you've got eyes and you aren't using them, they're a liability. Eyes take resources to build and maintain, and are prone to getting injured and infected. Cave creatures born with atrophied eyes may thus find themselves at an advantage and end up producing more offspring than their peers, helping the genes for atrophied eyes spread in the population.

With our acanthocephalans, evolution has worked in the other direction, building extreme complexity, seemingly to the detriment of the worms. But natural selection abhors bad ideas. It's the adaptations that better suit an organism to its niche and help it make babies that get passed down through the generations. So for the acanthocephalans, it was probably that some individuals had the lucky mutation that let them survive in the guts of newly arrived birds. Eventually the species adapted to not only survive there, but breed there, and indeed require the birds to complete their life cycle. This complexity is what encouraged the development of zombification, which in turn made the system all the more complex.

Same goes for *Ophio*. The intricacy of the ant colony is probably what led the fungus to develop such a mind-numbingly complex life cycle of its own. It couldn't kill its hosts in the nests, or it'd be escorted out. The parasite's solution, then, was to up the stakes by manipulating ants to its own ends. Same goes for the jewel wasp. Few things in the animal kingdom are more complex than the brain, and even fewer are more complex than brain surgery. But to ensure her kids survive in a cruel world, the mother wasp goes above and beyond, because long ago her ancestors stumbled upon a solution that natural selection selected for. It wasn't some watchmaker in the sky who dreamed up such a convoluted scheme, but the simple, elegant processes of evolution.

## Second Curse, Same as the First

Here's where things get even stranger. The acanthocephalans aren't the only worms that have taken notice of amphipods, aka the punching bags of the crustacean family. All kinds of other worms invade their bodies, and even though the parasites are unrelated, they can pull off similar manipulations. I mean, if a watchmaker were behind all of this, they'd have to be a very uncreative watchmaker indeed.

Consider the trematode worm *Microphallus* (yes, meaning "tiny penis"—glad you asked) *papillorobustus*, which plays out an even more complex life cycle than our acanthocephalans. It adds one extra stop, moving from a snail to an amphipod to a bird and back to a snail by way of droppings. But like an acanthocephalan, *Microphallus* steers its amphipod into the light and toward the surface, where birds await. So acanthocephalans and trematodes are brainwashing crustaceans to the same ends. What are the odds? And what are the odds they're going about it the same way? Well, pretty good, as it happens. And it seems to be a behavioral trick you're probably familiar with: the neurotransmitter serotonin.

Open up an amphipod laden with an acanthocephalan or *Microphallus* worm and you'll find its brain is drowning in serotonin. More intriguing still, a manipulative worm that ends up in its typical species of amphipod will set off a spike in serotonin, but put the same worm in the wrong species of amphipod—which it's unable to manipulate—and serotonin levels remain normal. For instance, in central France, our friend *Pomphorhynchus laevis* body-snatches the local amphipod that it's evolved to exploit. But it can't manipulate an invasive species of amphipod, because parasite and host aren't evolutionarily intertwined, so serotonin levels stay flat here. Pretty damning evidence that the zombifying worms are somehow futzing with the neurotransmitter.

Serotonin serves many functions in humans, most famously in regard to depression and anxiety (antidepressants like Prozac are selective serotonin reuptake inhibitors, meaning they stop the brain from reabsorbing serotonin, so more of the stuff can hang around). Yet you'll find serotonin across the animal kingdom, making it an ancient evolutionary invention that came about before many species went our separate ways on the tree of life. Its uses vary dramatically, though: In crustaceans like amphipods, serotonin helps regulate locomotion and the senses. Importantly for our mystery of the zombifying worms, sight is included among those senses.

But it's not likely that the worms themselves are dumping all of this serotonin into their hosts. Instead, the amphipod's immune system may be inducing inflammation to fight the parasite, leading to a massive release of serotonin. Overwhelmed, the brain begins to short-circuit (much like spikes of octopamine and cytokine make *Cotesia*-infected caterpillars go nuts), misinterpreting stimuli from the environment, including light. Thus scientists reckon the body-snatched amphipod may see light as darkness and darkness as light, so when it thinks it's fleeing down into the depths it's actually headed toward the sky. Indeed, if you inject an amphipod with serotonin it starts acting exactly as if it had a manipulative worm in its belly, rushing madly toward light when threatened. The zombie crustaceans that made a break for Bethel's and Holmes's leg hairs, then, may have thought they were digging into the safety of mud. So it would seem the zombifier is flipping the amphipod's world upside down not by evolving to inject the mind with serotonin, but by turning the crustacean's body against itself.

Here we have two unrelated worms that manipulate amphipods not only in the same way, but by the same means, glitching pure biology to get themselves into the belly of a bird. They've independently evolved life cycles that seem too intricate to be true, yet clearly it's working. I mean, if I were the watchmaker, I'd want to at least change things up a little bit.

Regardless of how this sort of zombification evolved, what's important to consider is that these parasites are splitting populations into two distinct subpopulations. Down at the bottom of the lake, in the darkness, tucked away from the predatory birds and fish swimming above, are the uninfected amphipods. Up top are the zombies, gathered up in hordes, as zombies are wont to do—darting around in the light, defying their own best interests. The two populations are segregated: Zombies tend to mate with zombies and the normals tend to mate with normals. They live in the same lake, but they may as well be different species.*

Which got me thinking: A manipulative worm gets its host eaten, which has implications for the host, obvious enough. But what about the environment as a whole? Zombie amphipods are pretty much handouts for predators, after all, yet surely the

---

*Consider as an analogy the fate of the human race as H. G. Wells imagined it in *The Time Machine*. Split between living as typical aboveground people and adopting a subterranean lifestyle, we evolve into two distinct species. The Eloi are more or less like us, in the sense that they're pleasant yet lazy, while the underground Morlocks are murderous brutes, making them . . . more or less like us. Okay, I see what H. G. did there.

manipulators' impact doesn't stop there. (I mean, I've never seen a zombie movie in which one person is bitten and that's that. There's what you might call a cascading effect.) And indeed, nature's real-life zombifiers aren't just influencing their food chains—entire ecosystems are built around them. This is far, far more complex than a worm and its crustacean punching bag.

This is zombification on a vast scale.

# 4

# No Creature Lives in a Vacuum, Not Even a Zombie

***Sure, go ahead, manipulate another animal into doing your bidding. But know that the consequences could ripple through an entire ecosystem.***

May I introduce you to hell—in frozen form: winter on Isle Royale, a smear of fragmented land painted like a brushstroke in Lake Superior, right off the coast of Wisconsin. Here you can find almost zero food, at least if you're a moose. Here you lose weight day by day, sometimes not even bothering feeding because it takes too much energy to struggle through the deep snow. Here if the cold doesn't get you, accidentally plummeting off a cliff will. And if that doesn't do the trick, a pack of wolves might find you and, over the course of several hours, nip and gnaw at your scrawny legs until the combination of that awfulness and the cold snatch you away.

Such is the life of a towering herbivore on Isle Royale. Here dutiful ecologists have been braving both wolf and moose to monitor both wolf and moose since the late 1950s, studying how the two species are fatefully linked. How predator and prey—the moose having waded over around the turn of the twentieth century, the wolves following when an ice bridge formed in the winter of '48—now find themselves trapped together. This is the interplay of species at its most dramatic.

And even here, the manipulators make their mark. A tapeworm known as *Echinococcus granulosus* lodges in the organs of the moose, particularly the lungs, forming gnarly cysts. When a wolf consumes these organs, it consumes the worms. The parasites take up residence in the predator's gut, releasing their eggs with the host's feces. A grazing moose inevitably eats a bit of scat (I repeat: life is rough for any organism, but particularly so for a moose on Isle Royale), and the whole cycle starts anew.

A moose with a lungful of worms has a doozy of a time

breathing. It slows down because it can't get enough oxygen to power its muscles, a particularly pressing problem when it's already sapped of energy in the lean times of winter. And even in the warmth of summer an infected moose is screwed, for like a dog it must pant to cool down its body. So no matter the time of year, an infected moose is a sluggish moose, and a sluggish moose is wolf food. Thus, by lodging in the lungs, the tapeworm may be pulling off a subtle manipulation to drive its host into the jaws of a wolf pack. No need to mess with the brain when all you have to do is cut your vehicle's supply of oxygen.

The most striking eyewitness account of this sort of manipulation comes from Alaska in the 1950s. Scientists caught sight of five wolf puppies chasing a pair of two-year-old caribou, one of which made a quick getaway. "The other ran," the researchers wrote, "but remarkably slowly." And so the pups locked on. Yet instead of picking up the pace, the caribou did the unadvisable: "It turned and faced them and sank on one knee, then lay down. Hesitatingly the small wolves surrounded it. It got up and ran. Then, untouched, it faced and voluntarily lay down again." The scientists later dissected the caribou, finding eight abscesses in its lungs, "some as big as ping-pong balls, half buried in the tissue and full of watery fluid." Worms.

That, however, is anecdotal evidence that the parasite's ravaging of the lungs makes large herbivores more vulnerable to predators. Luckily, we've got solid data from elsewhere to make the link: southwestern Quebec. Here scientists plotted three areas of low, medium, and high rates of wolf predation on moose, then asked hunters in each if they wouldn't terribly mind handing over the lungs of their quarries, please and thank you. The researchers discovered that in areas of more frequent predation, you find fewer infected individuals compared to areas of rarer predation, and among those individuals, you find fewer worms. The idea is that if infected moose are more vulnerable to predation, wolf packs will pick them off preferentially to uninfected moose. If you've got a heavily infected moose population, wolves will trim it down more than a healthier population. So the data lends good evidence that the tapeworm dramatically impedes the survival of its poor devil of a host.

In a place like Isle Royale, the ecological implications of a manipulative parasite like this are huge. The moose and wolves are stuck on this island together: The moose have no escape unless they brave a swim to the mainland, and the wolves have no escape *and* no other significant source of food.* Not only that, moose aren't exactly pushovers—they can and do stomp wolves to death, or shatter bones with a glancing blow, leading to a much more grueling demise. So a target that can hardly breathe is that much easier to take down. Indeed, the wolves of Isle Royale may *rely* on the tapeworm as a partner to make a decent living hunting moose. A decent living means healthier packs, which means fewer moose. Fewer moose means fewer grazers, and therefore more vegetation. (Before the wolves arrived, overgrazing was common, as was the consequent starvation of moose.) The tapeworm, therefore, transforms not only its host, but the very fabric of an ecosystem.

And it's not alone. The zombifiers aren't just manipulating their victims in astonishing ways, but in so doing are also manipulating their environment and the relationships of the species within

*Whenever you get a population as isolated as this, inbreeding becomes a serious problem. This is the fate of the wolves of Isle Royale. As of this writing, only two individuals remain on the island—a father and daughter that are also half siblings. They're so inbred that this appears to be the end of their genetic line, which means the end of the Isle Royale wolves.

it. After all, at the end of any good zombie outbreak, the world never looks quite the same.

## Parasites Warm the Heart of My Cockle

Scientific discoveries come about by way of raw grit and determination and obsession—think Einstein constructing the theory of relativity or Marie Curie unraveling the mysteries of radiation with such dedication that the poisoning killed her. Other discoveries, though, are rather more . . . fortuitous: Every so often someone literally serves them to you on a platter.

Take Robert Poulin, the man who practically invented zombie ecology. In 1997 he joined his postdoc, Fred Thomas, for a dinner of local New Zealand cockles, a kind of clam.* But when they cut into one cockle's muscular "foot"—which, if we're being honest, looks more like a toe, but whatever—they found something peculiar. "It looked like little grains of salt embedded within the flesh," Poulin tells me. "And we just looked at it, put it in front of the window. It looked to us like they were the encysted stages of juvenile trematodes that you find commonly in fish or in small invertebrates." Mental note made, the dinner party continued unfazed. "The cockles were cooked, so we were not too worried about it. We ate them."

The next day, Thomas went back to the market and got the seafood manager to hand over the name of the supplier. "We contacted them right after to ask where they got their cockles, and

*"Cockle" coming from the French *coquille*, meaning shell (*coquiller* being a blister that forms when bread cooks). The origin of the phrase "warm the cockles of my heart" is unclear, though it may have to do with the shape of the bivalve resembling that of the organ. Or, if it's more convenient, we can just blame it all on the French.

they said they got them from a local inlet, Blueskin Bay," Poulin says. The duo took a few more bivalves from the supplier, then inspected other bays nearby. Sure enough, these ecosystems were all lousy with infected cockles.

New Zealand's calm, shallow bays would be nothing without cockles, worm riddled or otherwise. The muck can be so packed with the things—perhaps twenty-five cockles in a square foot—that they'll form the literal foundation of a bustling ecosystem, at times the only hard surface other creatures can attach to. Two particularly important species grow on these cockles' shells: the anemone and the limpet (those ribbed, conical shells you've probably seen stuck to rocks in tide pools). And wading through these waters is a bird called the pied oystercatcher, which, true to its name, hunts shellfish. But that's not a problem for a clam hidden just below the muddy surface.

Unless, that is, a trematode worm has taken command of the cockle's body. The clam typically uses its muscular foot to dig down into the muck, from time to time resurfacing and crawling a few inches and digging once more. But what Poulin noticed at that fateful dinner party was that the worms had gathered in only one bit of the cockle. "The location of the parasite was very specific," Poulin says. "It was in the foot tissue of the cockle, not in other muscles, just specifically in the foot. And that immediately suggested that they may impair the burrowing ability of the bivalves." Indeed, by encysting in the foot, the worms seize it up and stunt its growth. Thus unable to burrow, the clam dislodges to the surface of the mud and sits exposed, making it easy pickings for human shellfish hawkers or, more commonly, the pied oystercatcher. And it's in this bird's belly where—surprise—the worm completes its life cycle.

But what Poulin has found is that the worm is doing far more than commanding its host—it's commanding the ecosystem. The anemones and limpets that attach to uninfected cockles typically live in harmony (though, full disclosure, an anemone may sometimes avail itself of limpet flesh), the former snagging tiny critters floating by and the latter grazing on the algae that grow on the host. But unlike the anemone, the limpet sports its own shell covering its body, so as the tide pulls out and exposes the manipulated cockle to air, the armored limpets have an advantage, growing more numerous than the fleshy anemones, which desiccate.

What we find, then, is that by meddling with the cockle foundation of the ecosystem, the trematode is throwing off the balance of other species. In a good way, you might argue: Overall, species diversity climbs as more body-snatched cockles pop out of the mud. Microhabitats form in between the bivalves, providing shelter for other species. More marine worms, more crustaceans, more surfaces for algae—which means more food for the vegetarians. More zombie cockles leads to more zombifying trematode larvae floating about, providing a reliable source of food for carnivores, including our anemones, which will grow *more* abundant if they attach to cockles in deeper parts of the mudflat that don't dry out at low tide.

But not everyone in the ecosystem profits from the trematode's exploits, because healthy cockles have a job to do, mixing up the mud with their forays to the surface. That blends the different materials that make up the sediment, from coarse sand to fine silt. If the cockles get infected and seize up, that sand begins to accumulate as a top layer, and some tiny invertebrates may not be able to tunnel into it, thus sitting exposed. That and the chemical makeup of the mud begins to change.

"As you go deeper and deeper, sediment becomes black—it

stinks—and very few organisms will burrow into that," Poulin says. "And the depth at which this appears, this anoxic zone, depends a lot on the biologic activities of those organisms that mix the sediment. So the more mixing, the deeper the anoxic layer begins." That means more top layers for critters to inhabit.

While not all invertebrates appreciate the trematodes, fish certainly do—which is problematic for the worms. The so-called spotty (really, it only has one blotch on either side of its body, so the name spotty*ish* would be more accurate) will make its way through the fields of stranded cockles and nip at the tips of their feet. But a fish gut is a dead end for the trematode, which is banking on a bird gut instead. In fact, 17 percent of the larvae are lost to spotties, while 2.5 percent make it into oystercatchers. Those seem like terrible odds for the parasite, I know. But this appears to be a high-risk, high-reward kind of manipulation, for oystercatchers in some mudflats feed nearly exclusively on stranded cockles. With such high rates of predation, the worm's strategy pays off.

Curiously, by nipping at just the foot, the spotty might be helping the cockle survive: The fish is essentially amputating the bivalve to cure it of its infection. The cockle can regenerate that tissue in two or three weeks, at which point it once again wields a functional foot, allowing it to squirm back into the mud. Eventually, though, it'll suck in more trematode larvae, which will seize up the foot. Having lost its anchor, the cockle will rise to the surface, where another spotty will amputate it. The bivalves, then, may be locked on a helpless roller coaster of happy times in the mud and nightmarish paralysis at the surface—up and down, up and down, all at the mercy of a worm.

Hop over to the coral reefs of Hawaii and you'll find another trematode with oddly similar impacts on its ecosystem. This worm attacks the finger coral—which at the risk of ruining coral reefs for you, looks like hundreds and hundreds of fingers stretching skyward—lodging in the small fleshy bits known as polyps. (Most of the coral body you typically see, that hardened shell, is the polyps' exoskeleton.) This is unfortunate for the coral on a number of counts. For one, infection reduces the polyps' cache of photosynthetic algae, which the coral relies on to absorb energy. That means the coral grows half as quickly. Infected polyps also swell up big-time, meaning they can't retract into the safety of the exoskeleton. They, like the cockles, are stranded.

This is good news for any number of butterflyfish species that patrol the reef. Polyps are their favorite food, and swollen polyps are both conspicuous and defenseless. And so fish graze on the infected finger coral, ingesting the trematode larvae in the process. The adult worm lodges in the gallbladder and the fish poops out its eggs, which hatch into the swimming larval form, which invades mollusks, which expel the next larval form, which seeks out still more polyps. Again, sorry to ruin coral reefs for you.

Except the coral—the foundation of the reef ecosystem—might in a weird way be benefiting from this predation. Build a cage around an infected coral to keep the butterflyfish out, and the worm will run rampant, stunting the coral's growth. But let the fish in to trim the infected polyps and the coral will return to its normal growth rate, regenerating healthy new polyps (as a cockle would regenerate its foot) that can once more harbor photosynthetic algae for energy. Thus, even though it's ridden with a nasty worm, the coral can refresh itself over the years—many, many years. These things may live for centuries, in fact, constantly regenerating bits the butterflyfish amputate. Still, the fish are in a way responsible for the coral's plight: The worm needs them to complete its life cycle, so the predators are both the problem and the cure.

So the manipulators can transform the ecosystems they call home. Zombifier and zombie—be they the worm+coral or wasp+cockroach or fungus+ant—don't live in isolation with one another. They end up changing the world.

## You're off the Hook This Time, Butterflies

Whenever I get on a plane, I can't help but think, *This is the machine that will undo humanity.* This is the machine that will take Zombie Patient Zero from one side of the world to the other, which wouldn't have been possible when the world's peoples were isolated thousands of years ago. Now we're one big family, which is great for interconnectedness and all that but also for the hypothetical spread of a virus that zombifies humanity. All it would take is a single person and a plane ticket to set off a cascade of misery the world over. (I'm a joy to fly with, I swear.) One little glitch in the system, and the whole thing comes crashing down.

Consider the famous butterfly effect (the theory is famous, not the butterfly): If a butterfly flaps its wings in Costa Rica, will it set off a tornado in Nebraska? Meaning, does the infinitesimal pressure that propels the insect change the surrounding air pressure just enough that it initiates a chain of atmospheric events that culminate in a twister touching down in Omaha? Well, no, sorry. The idea behind the butterfly effect—that even a tiny change in a system can have cascading, perhaps disastrous impacts far removed from the source—plays out in other systems, but it doesn't actually work out in this case. There's no way a butterfly could cause so much trouble. The principle, though, is key if we're to fully come to terms with what the zombifiers are ecologically capable of. Because while a butterfly flapping around is harmless, armies of worms attacking cockles and fungi manipulating ants and wasps slaughtering caterpillars most certainly are not.

I for one haven't the slightest clue what it's like to be part of an ecosystem. I know I share an apartment with microbes and insects and a cat—which wasn't my idea, for the record—but that's about it. We humans long ago excused ourselves from the food chain, opting instead to harness the power of agriculture and animal husbandry to feed our species, which has consequently ballooned to worrisome proportions. Free from the constraints of an ecosystem, we have thrown pretty much every other ecosystem into chaos. The species we've killed off, the forests we've decimated, the seas we've overfished. Even seemingly untouchable habitats like deep-sea vents are threatened by mining for precious minerals. And we have little clue how our biggest screwup, human-induced climate change, will transform the planet in ways both subtle and profound.

What science *does* know is that an ecosystem is a deceptively elaborate network of organisms that interact with each other and their environment in an astonishing number of ways, a place where the interconnectedness of life is on full display. Think about the zombie cockle mudflat. The trematode worms and cockles and spotties and oystercatchers are the critters you can actually observe. Poulin can quantify roughly how many cockles are infected in a bay, and how many fish and birds take advantage of them. That's the interaction of a handful of players. Then Poulin can show how stranded cockles provide new environments for the ecosystem's many other players. Straightforward enough.

But Poulin is one man, and this is a network that defies human understanding. What about the countless microbes like bacteria and viruses floating around the mudflat? What about the larvae of critters other than the zombifying worms? Understanding how a trematode's shenanigans affect individual species is, maddeningly, a tiny inroad into understanding the parasite's overall impact. This species over here may interact with that species over there, but not this other, but maybe this other, but only if another species isn't around. Oh, also, that may all depend on the weather, and the acidity and temperature of the water, and the time of year. The ecosystem has so many variables that you'd be a damned fool to try to map exactly how it's arranged.

The impenetrability of the ecosystem for the human mind becomes all the more problematic when the human hand introduces something that's not supposed to be there. Invasive species, whether humans bring them into an environment intentionally or accidentally, can have consequences that ripple through an ecosystem. Take, for instance, the rat, the most prolific invasive species on Earth. Humans and their ships have introduced it to islands all over the world, sending ecosystems into chaos. Islands that have never before seen a mammal (other than bats, which are quite frankly everywhere, but mostly keep to eating bugs and fruits) now find one obliterating bird populations by devouring their eggs. Without birds to hunt them, insect populations go haywire. Plants lose an ally, as there are no birds to eat their fruits and spread their seeds. And so a single misplaced species can set off bedlam.

Humans understand this power well. We can and do use invasive species as weapons, included among them the zombifiers. For instance *Cotesia glomerata*, cousin of the *Cotesia congregata* we've already met, attacks the cabbageworm, itself an invasive species that landed in the United States in the late 1800s. Shortly thereafter, realizing the worm had a bit of an appetite, officials imported *Cotesia glomerata*, pitting the two invasive species against one another. It kind of worked. The wasp only ever established in isolated pockets, where it doesn't have that big of an impact on cabbageworm numbers, even though the parasite specializes in destroying these caterpillars specifically. Which only proves the point further: You'd be a damned fool to pretend to know exactly how an ecosystem is going to play out.

Consider, too, our fungi and other wasps, and the ecological questions they raise. How well does *Cotesia congregata* control populations of the tobacco hornworm, also a pest? How would an ecosystem transform and crops suffer (unfortunately for gardeners, the tobacco hornworm attacks tomatoes as well) were it to disappear? And what about the fungus that invades cicadas and disintegrates their bodies? You might assume its disappearance would make those periodic cicadas even more insufferable, but it's impossible to quantify its impact on the insects short of multitudinous researchers scouring swaths of forest for fungus-eaten carcasses. As for *Ophio*, remember that its infection of the colony is chronic, so the ant society is forever struggling to contain it. How badly does the fungus handicap the colony, and then in turn, how does this affect the ants' impact on the larger ecosystem?

It's easy to get lost in the majesty of the zombifiers. We can sit in the basement of an Israeli university and marvel as a jewel wasp lobotomizes a cockroach and leads it to its doom. But out in nature, as we speak, the wasps are waging war on populations of roaches, altering the ecosystems they call home. And not just as we speak—they've been doing so for millennia upon millennia. As long as complex life has walked and slithered and swum the

Earth, you can bet the zombifiers have been right there with it, dramatically shaping the animal kingdom.

## On the Road with Hitchhikers, Lucky Passengers, and Copilots

So, we know that zombifiers like the cockle trematodes can have dramatic impacts on an ecosystem at large. But something even subtler is going on here: What happens when more than one parasite invades a single host, in a sense creating a tiny contained ecosystem?

Plenty of other parasites are more than happy to exploit the millennia of serendipitous genetic mutation after serendipitous genetic mutation it's taken the zombifiers to assume control of their hosts. These are the "hitchhikers," as Robert Poulin calls them—the opportunists along for a ride. "Certainly there are cases where two parasites have the same definitive host and the same intermediate host," he says. "But only one of them manipulates the intermediate host, which suggests that the other one is benefiting without doing any of the work."

The amphipods, for instance, don't only play host to zombifying *Microphallus* worms, but to another parasitic worm called *Maritrema subdolum*, which can't influence behavior one bit. The two worms have similar life cycles, though, moving from an aquatic snail to the amphipod to a bird, then back to the snail when the mollusk eats bird droppings. Their interests are therefore the same: Get into a bird at all costs. So while *Microphallus* lodges in the amphipod's brain and begins manipulating its behavior, *Maritrema* instead lodges in the critter's abdomen and bides its time. The manipulator worm drives the amphipod toward the surface, where, on a good day, a bird picks it off. So

*Microphallus* is the driver and *Maritrema* the mooch. It's a clever strategy for the former and an even more exploitative one for the latter—a hitchhiker of a driver of an unwitting vehicle.

But maybe *Maritrema* ended up randomly in the same amphipod as the body-snatching *Microphallus*, right? Maybe it isn't so much strategy as it is luck. Well, it actually appears that *Maritrema* is "seeking out" zombified amphipods, because the two worms show up together in hosts far more frequently than by chance, one study found. And that may be because hitchhiker larvae are almost twice as likely to swim up into the water column as the body-snatching larvae. Here, they're more likely to happen upon a suicidal amphipod with *Microphallus* driver worms already in its head. Sure, that puts the swimming larvae in danger of ending up in the stomach of any number of predators up there at the surface, but it also boosts their chances of ending up in the stomach of a bird by way of a zombie amphipod.

That may well be worth it, especially considering that while inside the host, the zombifying worm is doing all the work, expending the energy to manipulate the vehicle. The backseat driver can relax in the amphipod's abdomen and enjoy the free ride and food. Natural selection would favor this kind of opportunism, since the cheater can invest the saved energy elsewhere, like growing larger or producing more offspring. Indeed, it could be that the hitchhiking worm needs to invest that energy in rapid growth. Because it's not the one in control of the amphipod, it needs to reach its infective stage quickly, before a bird snags the host. If not, ending up in a bird intestine prematurely ain't gonna do it no good nohow.

Even if a parasite isn't actively "seeking out" prezombified hosts, the parasites in a given ecosystem could stand to gain from the exploits of the body snatchers. These aren't hitchhikers, but what Poulin calls "lucky passengers." For instance, the same species of manipulative worm, *Microphallus papillorobustus*, will at times share an amphipod with a nonmanipulative worm, *Microphallus hoffmanni*.* Once again, both parasites need to end up in birds. But unlike *Maritrema* hitchhikers targeting zombified amphipods, it seems this second variety of nonmanipulative worm isn't necessarily squirming up into the water column in pursuit of hosts. It gets to a manipulated amphipod on a more random basis—riding upwelling currents, perhaps. Still, that chance encounter will boost the lucky passenger's chances of arriving in a bird gut. So *Microphallus papillorobustus* makes a far greater impact on its ecosystem than simply ruining the day of an amphipod: It provides food (and parasites) to birds and helps further the exploits of at least two other worms.

Manipulative worms also face competition for control of their vehicles, a particular problem when two worms use the same amphipod to get into different definitive hosts. Take, for example, *Pomphorhynchus laevis* (the color changer we met last chapter), which needs to get into a fish, and *Polymorphus minutus* (the one we met that speeds up its host), which needs to get into a bird. When alone in an amphipod, the former will steer its host out of shady parts of the pond and into the light, where a fish is more likely to take notice. The latter will steer its amphipod to the surface, where birds prowl. But get both worms in one amphipod and you've got yourself a showdown of competing interests. There's no "halfway" that the two worms can agree on, no compromise between the pursuits of light and depth—short of

*Which translates to "tiny penis Hoffmann." Who Hoffmann was, and why the scientist who named the species after him in 1964 would do such a thing, is a mystery.

splitting the amphipod in two and trying their luck that way. And indeed the fish-bound worm seems to grab the steering wheel in this scenario, driving the amphipod into the light, while the bird-loving worm isn't able to induce the amphipod to swim as high up in the water column as it would if it were alone in the host. So like two cults fighting to brainwash one follower, one parasite is able to wrest control of the host from another.

This is a fascinating manifestation of what's known as coevolution. Zombifying worms, or any species for that matter, aren't evolving in a vacuum. They're evolving alongside (or inside, really) their hosts, at some point developing an exploitation, which the host might counter by evolving a defense, which the worm in turn may find a workaround for. Each party is putting pressure on the other. Same goes for two species of manipulative worm that fight for control of a single host species. These parasites have to evolve with their host to overcome its defenses *and* they have to deal with one another. It's not like they have mouths with which to bite each other to death, so they've had to evolve better and better manipulative strategies to gain the upper hand, trapped together in the body of a crustacean.

Conversely, when *Pomphorhynchus laevis* shares a definitive host with another worm, as is the case with *Acanthocephalus clavula*, they appear to have coevolved to share manipulative responsibilities. Once again, in this scenario the worms are far more likely to show up in the same amphipod together than they would by chance, suggesting there's some sort of evolutionary benefit to associating with each other. And once again, *Pomphorhynchus laevis* is able to steer the amphipod into the light. Same with *Acanthocephalus clavula*. But the former can also change the color of the host to make it all the more conspicuous. Thus the two manipulative parasites become "copilots," both of them changing the host's attitude while one changes its outfit to make the stooge stick out all the better.

This kind of cooperation could also be the case with the cockles of New Zealand. In 2004, a few years after that dinner party, Poulin discovered that the bivalves harbor not one but two different trematodes, the new species being about half the size of the other. A few years after that, genetic work revealed that those two species are actually six distinct species that will take up residence in a single cockle. All must find their way into an avian belly, so all encyst in the tip of the cockle's foot.

"It looks like it's teamwork," Poulin says. They're sharing the responsibility of manipulating the host to get where they need to be. They aren't brainwashing the cockle like their wormy comrades do with amphipods, but their combined damage to the foot forces their host out of the mud and into the open, where, ideally, the six species of trematode will end up as a package deal inside a bird.

And so the body snatchers aren't just altering the environment at large—driving unfortunate critters into mouths and in some cases functioning as the foundation of a whole ecosystem—they are creating unique communities *within* their hosts. Worms battling worms, worms partnering with worms, worms taking advantage of other worms' accomplishments. Each zombie isn't a mere vehicle, but a collective of manipulators and nonmanipulators alike, where complex life cycles collide to create something exponentially more complex.

Now, if you're anything like me, by this point you're wondering what it's like living inside a cockle or amphipod. What it's like to cozy up in the warmth and darkness, maybe with a friend or two, kicking back as your host ferries you around. Free meals, free rent—the good life. Except it turns out the existence of a parasite is anything but easy. In fact, it's a special kind of hell that rivals the zombie's own travails. And to prove it, I hiked into the hills of New Mexico, where I was promised there probably weren't any cougars or bears.

# 5

# How to Succeed in Parasitism Without Really Dying

***So you want to brainwash another creature to bend to your will. First, though, take a lesson from the nematomorphs: There's nothing simple about living in the belly of the beast.***

There's signs about cougars and bears," Ben Hanelt mentions, elbow stuck out the window of his SUV. "But just ignore them. I've never seen cougars or bears up here."

We're an hour south of Albuquerque—where it isn't cacti that reign, but pines and oaks—firing up dirt roads flanked by trees turning orange and red and yellow. It's early fall, but they call this place the Fourth of July—you know, on account of the colors. I call it something else: the Land of the Nematomorph Zombifying Worm That Invades Crickets and Mind-Controls Them into Jumping in Water, at Which Point the Worm Erupts Assertively from Its Host's Abdomen.

As we park and pile out of the truck I can't help but think, Well, yeah, but Ben, *someone* must have seen at least one cougar and one bear to warrant the sign. Which is why I'm happy to let Mr. Hanelt—dressed in flannel and backpack and neatly trimmed beard—lead this hike up a dry creek bed in search of parasites, holding a blue water bottle with which to imprison them.

As we trudge, the only sounds are our footfalls on the leaf litter and that *whooosh* of wind slicing through pines. And then we come upon a tremendous shit on a rock.

"That's a biggun," I say.

"That kind of looks like bear. Oh *yeah*, maybe he ate *him*," Hanelt says, pointing at a red shirt lying nearby. That'd be red with dye from the factory, by the way, not human blood. I like this Ben Hanelt.

Hanelt continues up the creek, lamenting that it's dry—*very* dry. So dry, in fact, that when we reach the spring he was banking on, well, springing forth water, it's doing no such thing. That means no nematomorph worms here—but on the upside, no bears or cougars either. "What we'll do on the way down is look in the more moist areas," Hanelt says as he clambers over rocks. "A lot of times these worms will actually dig down into the mud."

A hundred feet or so down the hill Hanelt stops, bends down, and flips over a slab. Stuck to the underside are three little tubes of pasted-together pebbles. Caddis flies. More specifically, the homes that caddis fly larvae assemble with gluelike spit to protect their squishy bodies. This life may be the wrong life as far as Hanelt and I are concerned, but it's a good sign, for nematomorphs invade the bodies of caddis fly larvae too.

And sure enough, there in the mud below the uplifted rock lies a perished nematomorph. It's but a strand—cream colored and exceedingly thin, curled up on itself. Unfurled, it's probably eight or nine inches. Not too shabby, and especially not too shabby for a worm that at some point in the recent past erupted out of the abdomen of a cricket a fraction of its length.

"Yeah," Hanelt says, dropping the caddis fly larvae in his bottle. "Let's go look for some live ones."

And so we make our way through the *whooosh* of the forest to the truck and ramble a quarter mile down the road and pile out once more. Now we can hear running water. Not a gush, but a trickle. We duck through some brush, over a barely running stream, and into a clearing abutting a shallow cliff—a lovely scene where you might expect a unicorn to frolic if such a thing had a role in the natural history of New Mexico. And here is what he was after: a circular cattle trough the size of an inflatable kiddie pool.

"Oho!" Hanelt yells. "Hallelujah! Look . . . at . . . *this*. They're happy little critters!"

In the tank is a grand assembly of the nematomorphs, two

dozen of them—thin and wriggling, with tan bodies for the ladies and darker bodies for the boys—squirming along the silt that coats the bottom of the trough. We know crickets must have brought them here, but we can't find a single drowned insect. Instead, on the surface water bugs jet about and yellow oak leaves drift. Through these Hanelt plunges his hand to grab the worms. "*Maaan*, that water's cold." One by one he piles them in his palm, forming a wriggling knot. "Look at that." It's clear now why folks also call these Gordian worms, after the Greek myth of Alexander the Great untying the legendary Gordian knot by cleaving it with his sword, which is cheating.

Hanelt asks me if I want to hold the knot, and of course I do, so he drops it in my palm. The whole mess isn't squishy like you'd assume worms to be. Instead, they feel like al dente angel-hair pasta. "They're literally tubes of gonads with tiny little muscles, but you can feel them sort of wriggling around on you," Hanelt says as he plucks more worms from the water. "For what tiny little muscles they have, they can actually put on quite the show."

Hanelt tells me to throw my Gordian knot in, so I do, as he dumps others into his water bottle. Satisfied, we walk back through mushy grass to the SUV. Hanelt drops his friends into the cup holder, starts the engine, and gets us rolling back to his lab in Albuquerque.

## I've Got a Gut Feeling

In a muggy room at the University of New Mexico, packed with racks of plastic tanks, themselves packed with snails, Hanelt and undergrad Rachel Swanteson-Franz cultivate nematomorph zombie worms. On account of some of the snails having tropical tendencies, the temperature is set to precisely 82 degrees

Fahrenheit (according to a comically large Flavor Flav–style thermometer hung from one of the racks). It smells like you'd expect over a hundred tanks of snails to smell at precisely 82 degrees.

Swanteson-Franz fills a Pyrex bowl with a finger or two of water, pulls a cricket from a bin, and drops it in. As soon as its host hits the water, the nematomorph makes its move, squirming sinuously out of a hole it's drilled in the cricket's belly. This appears to be not such a big deal for the cricket. The worm wriggles back and forth, growing longer and longer, but only once in the fifty seconds it takes the worm to emerge does the insect so much as twitch its legs. The rest of the ordeal it just floats there, as if a worm seven times its length *isn't* spilling out of its belly and swimming free. Even when the worm is at last liberated, the cricket simply floats—even as the whips of the nematomorph's long, thin body occasionally smack it around the bowl.

Swanteson-Franz isn't done. She drops another cricket in another Pyrex of water and sets the bowl next to the first. This cricket has more moxie. It struggles to swim away from the alien emerging from its gut, bouncing off the wall. But before the second subject even finishes with its ordeal, Swanteson-Franz adds another cricket-cum-worm to the mix.

This new subject, too, tries to flee the parasite it now finds itself birthing, as its compatriot in the bowl next door continues its own quest for relief. At last the second cricket's worm frees itself, followed not long after by the third. Everyone now rests. And while the crickets will never know it, in this lab they got off easy, for in nature their ordeal is far more precarious.

It was no accident Hanelt and I found so many nematomorphs in that cattle trough. (Nematomorphs are also known as horsehair worms, because ranchers will discover them wriggling in their waters.) Find a puddle or stream or even an abandoned cup in the hills of New Mexico and you're almost sure to find the worms. And indeed, across the United States and all around the planet—over three hundred nematomorph species tallied so far—people find them in dog bowls and, more troublingly, toilets. They'll stomp on a cricket, pick it up with a tissue, toss it in the commode, and flush. Sometime later the worm will make its way back up the pipes and the poor homeowner will email Hanelt in a panic, thinking a human in the family is producing these things, when in fact that's impossible—unless you're a family of crickets. "I could have become a rich man by selling some sort of sugar pill," Hanelt giggles. "Here, that's a really bad parasite, take this. It'll cost ya four hundred bucks. It'll cure you right away!"

What's really happening is a zombie attack of global proportions. When a nematomorph gets into a cricket, the insect doesn't act at all differently, even as the parasite feeds on its bodily fluids. Only when the worm grows to sexual maturity and decides it's had enough of this place do problems arise. Somehow it convinces the cricket to leap into the peril of a stream or pond. Because once the worm makes up its mind to leave, it punches a hole in the cricket's exoskeleton—peeking out just barely, but just enough—and in this way it can "taste" the water it seeks.

To test this, Hanelt came up with a brilliant experiment. He made cricket saline, mixing up a solution with the same salt content and chemicals that you'd find inside the insect, and dropped an infected cricket in. "The worms ultimately all came out very sluggishly," Hanelt says. "But they never made it, because they thought they were still inside the chamber of the cricket." But when he dropped this same cricket into regular old tap water, the worms in its belly tasted the difference and immediately erupted. And should you so desire, you can also go halfway and take a cricket and dunk it in water real quick. "If you then immediately pull it out, dab it off, the worms will go like spaghetti, *shleeerp*, straight back in," Hanelt says. "Because it's possible that the cricket is just moving around when it sprinkles or rains a little bit, and ultimately the worms need to have a mechanism to know that they're deep in water."

It's at this point I remember that the trough we found was filled with dozens of worms, yet somehow no crickets. Hanelt assures me that even being exposed to such trauma, and even though the worm can make up half its weight, a cricket may survive and escape watery doom. The things survive fine in Hanelt's lab, after all. In fact, even though crickets typically harbor one worm, Hanelt was once able to infect a single cricket with over thirty, all of which erupted en masse without killing their host. (Which, depending on your perspective on mercy killings, either weakens or strengthens my argument that nature is cruel.)

But a lab is no forest. What happened to those crickets? Did they merely snap out of it and scurry out of the trough? Or did birds make off with them? I ask because, again, we can't consider our mind controllers just in a vacuum, but in a larger ecological context. Free meals of zombie cricket hordes could well transform how local birds structure their diets. And while no fish swam in that trough, for certain finned predators in certain river systems, kamikaze crickets make up a significant portion of their diet.

This is particularly true in Japan. Here, one study found that in five separate mountain stream systems, more than half of the trout population had crickets in their bellies—and crickets aren't supposed to be in rivers unless something has told them to be there. In fact, in the fall, the trout catch more wormy crickets by mass than any other prey item. Meaning, first of all, clearly Japan's got a lot of little zombies hopping about. And two, these poor crickets are a tremendous resource for fish, to the point where the fish might require them to survive, like the Isle Royale wolves rely on worms to help bring down moose. So again, parasitic manipulations have consequences that ripple through the ecosystem.

Bring our view down from the macroscopic level to a microscopic one and we'll find that nematomorphs live a lifestyle—past all the business about wriggling out of other creatures' bodies—that couldn't be more different from our own. It's not just the brainwashing that's so bizarre, but an overall existence of peril locked in the prison of another animal, one in which the worm has to somehow extract enough energy from its host to keep itself alive, yet not so much that it drops the cricket dead. Because for as much cruelty as parasites dish out, they're also subjected to extreme tribulations.

## Insane in the Membrane (Insane in the Brain)

By virtue of being insects, crickets have an open circulatory system—the inside of the abdomen, known as the hemocoel, holds a sea of the insect version of blood, known as hemolymph, which coats the organs. This means something like a parasitic worm has wiggle room in there—and a lot to eat. Given that the nematomorph has no mouth with which to chew, it instead soaks up nutrients through its skin. But the worm won't contain itself in the abdomen: It'll snake one end of its body into the cricket's head. Even when Hanelt loads up some sad cricket with two dozen worms, only one—only ever one—will do this. Since someone's always got their tail in the head, whether a worm is alone in the cricket's body or with friends, we might hypothesize that this is how the nematomorph delivers its kamikaze-inducing chemicals.

But what kinds of chemicals could convince the cricket to risk its life by diving into water? Hanelt doesn't know that yet, but it's probably our old pals the neurotransmitters. And genetics could help him find the answer. As genomic testing gets cheaper and cheaper year by year, cash-strapped scientists burning through grants are getting better and better access. For Hanelt, now it's a matter of sequencing the nematomorph genome to pinpoint the genes that control behavior. "We've already found thirty, forty of them that are in other animals highly correlated with behavior," he says. "And then the next step in this is, okay, once we know what the dictionary is of all the different genes, let's see how those genes are expressed, meaning how those genes are turned on and off in various scenarios." For instance, genes could potentially turn on to produce neurotransmitters when the worm is ready for the cricket to take a dive.

Now, you're familiar with the mafia, yes? The fellas that stroll around the neighborhood, occasionally kissing each other's rings, threatening shopkeeps with physical harm lest they pay a fee? Well, I'd like to introduce you to the mafia hypothesis, the idea that hosts can either give in to parasites and escape relatively unharmed, or get far worse than having their shop burned to the ground.

In the case of the cricket, it will perish if it can't off-load its worms, which, removed from the water that sustains them, will perish as well. Everyone loses. "Or," Hanelt says, "you can ultimately be manipulated by the parasites to play along and say, 'Okay, I'm only going to have this worm for twenty-eight days. After that I cooperate and I carry the worm to water and let it go, then ultimately I will survive, too.'" Should the cricket make its leap into water, and should it avoid drowning or ending up in a fish's stomach, everyone wins.* Not giving in to the nematomorph's demands, however, is 100 percent fatal. So by playing along, at least the cricket has something of a chance.

But let's back up to the beginning of a nematomorph's life. A mother worm will stick her eggs—one after another in a string—to a rock or stick on a river bottom. And right as the larvae hatch, their lives get complicated. Anything with a mouth will eat them, like snails and mosquito larvae and our friends the caddis flies, too. But this was the worm's plan all along. Inside these bodies the worm larvae, which look like elbow pasta (the high number of pasta references in this chapter is strictly coincidental), burrow into tissue and encyst, growing a thick protective layer and essentially going into hibernation. Even when the mosquito or caddis fly larvae metamorphose into adults, the cyst goes along for the ride. Think about that for a moment: As the larva's tissues transform into something totally unrecognizable from the original form, somehow the cyst ends up still a cyst in the adult.

But in the end, the larvae don't give a damn about the mosquito or snail or caddis fly—they care about crickets. To get there, the cysts can move with ease through the ecosystem, bouncing between hosts. Hanelt will feed nematomorph-encysted snail flesh to a crustacean, and the cysts pop up there as well. Feed the crustacean flesh back to another snail, and the cysts will appear once

---

*Infected crickets will only ever make this potentially deadly journey at night. The first to figure out this timing of manipulation wasn't a scientist, but a pool cleaner in southern France. To his dismay, every morning he'd find thirty or forty worms wriggling in one particular pool, which was surrounded on all sides by forest. He made it his mission to stake out this body of water all day, and saw not a single nematomorph appear. But sure enough, the next morning, more worms—and more work for the pool cleaner. *Merde.*

more. Feed the snail to a midge or mayfly, and the cysts still go along for the ride. "I call this the pinball machine model," Hanelt says. "You basically go around from one host to the next to the next to the next until you actually get where you want to be."

Cue the crickets. When that midge or mayfly inevitably perishes, it crashes to the ground, and crickets aren't picky about what they eat—they'll gobble up those cyst-infested corpses without hesitation. And in their bellies the encysted worm larvae will roar to life, using needle-like mouthparts to drill through the digestive system and into the body cavity.

What makes the nematomorph cysts such bad news for crickets isn't just that the things work their way into so many potential prey items, but that the cysts don't need to be fresh to survive. They themselves are undead: Dry out a deceased, infected mayfly for a few years and feed it to a cricket, and sure enough the cysts will emerge in their new host. Pop the worms in the freezer, and you're really only making them mad. "We've taken them up to five years, but I'm sure it's indefinite at minus 70 degrees Celsius. They'll literally be fighting their way out of the ice as they're thawing out," Hanelt says. Dry them or freeze them, the things will not die.*

Nematomorph cysts are so hardy, in fact, that Hanelt figures one could bounce around an ecosystem for decades, which is odd in and of itself, not to mention from an evolutionary

---

*Adult nematomorphs, I'd like to note, aren't exactly pushovers either. Should a predator like a frog snag a cricket, the nematomorph within has about five minutes to worm its way out of the amphibian's stomach before dissolving in gastric juices. The frog clearly senses that the situation is less than optimal, smacking its face as the worm inches out of its mouth, flicking its tongue to reel the prey back in. Eventually the frog gives up, opening its maw so the spunky nematomorph can wriggle to freedom.

perspective. With life spans of mere months, these worms go through the generations quickly, which means they can evolve faster than something as long-lived as a human. The thing is, by bouncing around the ecosystem for so long, nematomorphs don't have the relatively uniform life span that humans do—some immediately get into crickets while others meander for years. "How does evolution then work?" Hanelt asks. "If evolution wants to take you in one direction, right, but here you keep getting these ancient genes from fifty years ago into the population?" Great questions that Hanelt is just beginning to consider, much less comprehend. I mean, imagine the evolutionary shock of a human from twenty thousand years ago suddenly showing up and jumping back into the gene pool.

## It Ain't Easy Being Mean

It all feels rather foreign, this parasite business. But here's what's really going to be hard for you to come to terms with: Scientists estimate that over half of all animal species on this planet are parasites. Which means you and I, we're the minorities. Statistically speaking, *we're* the freaks. Nematomorphs, murderous fungi, brain-surgeon wasps, they subscribe to the predominate MO of life on Earth. Predominance aside, however, their parasitic kind hasn't risen to power without running into a few obstacles. A bloodsucking horsefly must dodge the equine tail, after all, and bedbugs must find beds. While the body-snatching parasites we've met may seem cunningly equipped to get the better of their hosts, and may seem like freeloaders floating lazily in body cavities, in reality they're up against serious defenses—and even more serious odds.

To evolve into a parasite is to deal not only with the environment that is the world—the water and air and trees and such—but with a new *living* environment, "the first environment that will defend itself," as biologist Claude Combes notes in his classic work of parasitology *The Art of Being a Parasite*. Immune systems deftly seek out intruders, from microbes on up to wasp larvae and foot-long nematomorphs, quarantining and destroying them. And a parasite with multiple, totally unrelated hosts has to somehow evolve to resist distinct defenses. Everything is trying to kill it—the bodies it invades, not to mention the inhospitable environment outside the relative comfort of those bodies.

So what's a body snatcher to do? In the case of our nematomorphs, the solution may be a rather fortuitous one: The worms devour the most energy-rich bit of the cricket, its fat, which also happens to be the seat of the insect's immunity to invaders.

Now, a cricket's immune system is decidedly not human. Because its insides are essentially an ocean of hemolymph and guts, protecting the body from attack is a matter of the fat fortifying this ocean with two complementary defenses. First, it produces the hemocyte, an immune cell which, if you'll remember, *Cotesia*'s viral coconspirator targets. The fat also produces proteins known as AMPs—antimicrobial peptides—that "flag" bacteria for the hemocytes to seek and destroy. These AMPs also go on the hunt themselves, breaking down the proteins that make up the bacterium's cell wall. In this way the cricket is able to ward off everything from bacteria to viruses to larger parasites like worms.

That is, unless a nematomorph destroys the fat, crippling the cricket's immune response. When the larval worm drills out of the digestive system and into the body cavity, it makes its way into the fat stores and begins feeding. Not with its mouth, mind you—again, it has no such thing. So it's likely the nematomorph didn't need to evolve to target the fat in order to ensure the cricket's immune system doesn't kill it. The worm just goes after the most nutritious part, which is also the cricket's defense against parasites invading its body.

How the nematomorph is able to break down this fat without jaws, though, is unclear. Hanelt reckons the worm may be excreting some sort of enzyme that dissolves the tissue, allowing it to slurp up liquid nourishment through its skin. Also unclear is how the larval worm is able to elude an immune response before it can consume the fat and shut down the immune system. "Is it something about the cuticle, that the worms are covered in this stage with maybe some sort of acellular protein that simply cannot be recognized by the immune system of crickets?" Hanelt asks. "We have no idea. But we've never seen in, gosh, thirty years of studying these things, we've never seen an immune reaction by a cricket, or any other insect, really, to the developing nematomorph larvae."

Regardless, the worm is able to knock out the cricket's immune response. But in so doing it's both saved its own life and weakened the integrity of the vehicle it's relying on to survive. Without an immune system, the cricket can't mount a defense against other parasites trying to get in. And the cricket has enemies aplenty, which grow all the more menacing once the worm punches a hole in the exoskeleton and pokes out to taste for water. "So you basically have set up this beltway, if you will, the worm sticking out—bacteria, viruses, fungi, falling on this worm—and then retracting back in," Hanelt says. Once that armor is breached, the whole cricket opens up to infection, on account of that ocean of hemolymph swishing around. More parasites mean more competition for resources, and the very real possibility of the pathogens crippling the worm or its trusty steed.

So how does the cricket survive? It might be that the nematomorphs are somehow replacing the immune system they destroy, not unlike *Ophio* might be replacing portions of an ant's nervous system. Specifically, Hanelt had a hunch that the worms may be producing some sort of antimicrobial. To prove it, he took nematomorph-stuffed crickets and injected one group with bacteria and another, control group with cricket saline, then removed the worms and tested them to see which genes had ramped up to produce compounds that fight infection. "Lo and behold, a lot of these are actually antimicrobial peptides that are found in insects," Hanelt says. So the worm may be deploying these compounds to boost its host's immunity.

"What we want to find out is: Are the worms making these things and secreting them into their host, basically protecting their host along with themselves?" Hanelt asks. "In parasitology it would be a unique finding. Obviously it makes sense. I think evolutionarily, it's so intriguing. It would definitely sort of change the way we think about what parasites are capable of." It's a worm reaching way across the tree of life to destroy and assume the immune system of a cricket. (I'll say it again: The species of Earth may be distantly related, but they're all intimately entwined.) And once the host and parasite part ways, theoretically the cricket can heal its exoskeleton and start building up its fat reserves again, reconstructing an immune system once more.

The cyst stage of the nematomorph, however, is far less active in its thwarting of the host immune system, though in fairness, what it's able to pull off is still an incredible feat. The larva can encyst in a range of species, from snails to insects to crustaceans, which means it has to figure out how to avoid the distinct immune responses of each (yet still somehow recognize when it's in a cricket, and a cricket only, so it can emerge from the cyst and develop into a body-snatching adult). It's not likely, then, that the larva is tailoring its defense against each and every species, but instead is relying on a one-size-fits-all approach with a thick cyst wall—about as thick as the folded-up larva is wide. It's likely excreting this from two special glands, each containing a substance that is on its own ho-hum. But like an epoxy that only gets sticky when you mix two components together, the glands combine to create a unique compound that in turn builds an inert cyst wall, which the host's immune system passes over.

And that's really something: Put a foreign object in your body and the immune system goes nuts, attacking the intruder with defensive cells. But not with nematomorph cysts in bugs. "We don't see this surrounded by immune cells," Hanelt says. "The immune system just ignores it completely." Thus the nematomorph is the Trojan horse of the zombifying parasites, hunkering down patiently until it reaches the cricket, where it erupts from the cyst and assumes control of the target by wiping out its defenses. As they say, all is fair in parasitism and war.

Destroying the immune system of a host is precarious enough, but by devouring the fat stores of the cricket, the nematomorph is in danger of crippling its vehicle in another way: energetics. You and me, we need only worry about supplying ourselves with energy. But the worm has to strike a balance within the cricket—sap as much energy as it needs to grow into an adult, yet keep the host healthy enough to ferry it to water. And remember that the worm can make up half the weight of the cricket, so it's also weighing down its host, forcing it to expend more energy to get around.

Hanelt poses a thought experiment. Your job is to remove half the weight of a car—never mind why—yet still be able to drive it.

So you start by ripping off the doors. The windows have to go, too. Taillights and carpeting and all the seats save for the driver's. "Sooner or later you have to take parts of the engine out, right, because that's where most of the weight is," Hanelt says. "Well, which part of the engine are you going to take out without shutting the car off?" This is known as the host energetic resource constraint hypothesis, or HERC. If a parasite expects its host to not only survive but bend to its will, it has to allot the victim a certain amount of energy. At the same time, the parasite has to steal enough energy to grow into a reproductive adult.

The best illustration of HERC comes from our old friend *Dinocampus coccinellae*, the wasp that turns ladybugs into twitchy bodyguards. One clever study found a link between ladybug survival while guarding the developing pupa and the fecundity of resulting female wasps. Conveniently enough for the researchers, females emerge from their cocoons loaded with mature eggs. The scientists counted these and found that the fewer eggs a female wasp was born with, the longer the ladybug survived. The more eggs, the shorter the ladybug's miserable remaining time on Earth. It would seem, then, that the wasps run into an energetic trade-off parasitizing their hosts: If they suck more energy out of the ladybug and put it toward developing extra eggs, their victim is more poorly equipped to survive the bodyguard phase. It's the high-stakes gamble of variation, some individuals betting (by chance, of course) on a longer-lived bodyguard and fewer eggs, and others opting to wear out their bodyguard fast, but as a result pumping more offspring into the ecosystem. Both techniques have their advantages—just not for the ladybug, obviously.

## I've Got a Parasite, and the Only Prescription Is More Fever

So parasites have to manage their energy intake and expenditure while at the same time dodging their hosts' defenses, whether it be by masking their chemical signature or encysting or outright destroying the immune system. But some hosts have a countermeasure to fight their invaders that may be harder for parasites to overcome: They turn up the heat.

When you get a fever, your first instinct might be to get rid of it. Take a cold bath or an ibuprofen, think really hard about that scene in *Titanic* when the guy lets the lady have the piece of wood to float on in the freezing water while he turns into a handsome Popsicle. But you might want to rethink that strategy. A fever is a bummer, but it's also an indispensable weapon your body deploys to fight infection. The idea is that the human form can tolerate a range of temperatures, but the nasties inside it might not. So by turning up the thermostat, you can in a sense cook off things like viruses, giving your immune system a boost.

Our bodies do this automatically, whether we'd like them to or not, but certain creatures have a degree of control over their fevers. Recall *Entomophthora,* the fungal plague of the housefly. It develops rapidly in its host, driving it to an elevated position on the fifth or sixth day after infection. But on the second or third day, the flies—apparently "aware" that something is awry—will spend more time in the sun. This is known as behavioral fever, and it allows the flies to mount a defense against an otherwise lethal parasite. Indeed, when researchers exposed infected flies to higher temperatures on days two and three of infection, over 90 percent were able to kill off their parasites and survive. But warm up the flies on day four or five and it's too late—the fungus ends up overwhelming its host. Out in the wild this strategy is of course dependent on the weather: If the fly happens to pick a cloudy week to get infected, it's in trouble unless it can find some other heat source.

And turning up the heat doesn't come without costs. A fever demands extra calories and oxygen intake, a particularly tall order when the fly is already weakened and forced to waste time sitting around warming up instead of feeding. (The bigger the dose of fungus that researchers give the flies, the more time the flies will spend sunbathing. The insects' bodies somehow know how badly they're screwed in each case and adjust.) And by sitting out in the open, the host is asking to get eaten.

Now, remember that the *Entomophthora* fungus will best grow in relative coolness. That's why it's evolved to order the fly into an elevated death posture in the late afternoon, setting the scene for an evening of optimal spore development. And that manipulation is suspiciously familiar to what the fly already does with its behavioral fever when it's infected with other parasites. It may be, then, that the fungus is hijacking that normal behavior and turning it against the fly, driving its host to an elevated position once things have cooled down instead of heated up.

Because these manipulations are so precise, we might wonder what effect higher temperatures have on the development of such body snatchers. Fever can suppress the development of fungi and worms, so does that mean it also mucks up their ability to modify behavior? Could the strategy for the host be to raise the temperature early enough in the infection to ensure it doesn't end up brainwashed? After all, there's a threshold when an insect stops being an insect and starts being a puppet. Perhaps it's possible for the puppet, then, to snip its master's strings before it's too late.

Hosts will also turn *down* the thermostat to mount a defense against parasites. A bumblebee, for instance, is fiercely loyal to its colony and its queen, fighting and dying not to pass down its own genes, but that its breeding siblings may pass down theirs. Which complicates matters for the conopid fly, perhaps the deftest of the body snatchers. It will tackle a bumblebee in midair and, while tumbling, pry apart the abdomen segments of its victim with its curved, can-opener-like bum and inject an egg—all before the bee hits the ground. Should the egg hatch into a larva and develop in the bumblebee, it will eventually order its host to dig down into the soil. At this point the larva dispatches the bumblebee and pupates in the relative warmth as temperatures plummet aboveground.

The bumblebees, though, seem to have a countermeasure, because infected workers don't return to the warmth and safety of the nest at night. Instead, they stay in the fields. Researchers found that only 5 percent of the workers they collected from a nest at night were parasitized, compared to the workers they caught returning to the nest in the morning, of which 43 percent had devils in their bellies. So the bumblebees are trying to freeze their parasites to death out in the elements. And indeed, by keeping infected bees in cold temperatures, the researchers found that about half of the larvae were unable to make it to adulthood.

But the bumblebee isn't just camping out to keep the parasite developing within it from infiltrating the colony. No, it's more selfless than that: The bee is prolonging its life, however briefly, so it can keep working for the good of the family. By spending the night outside the hive, the bumblebee lowers its temperature and slows the development of the insidious larva it harbors. In the morning, it wakes up and goes about its foraging, providing its family with nourishment. The next night, while its comrades sleep at home, the bee again camps out in a field, lowering its temperature to stymie the parasite within. And then it wakes up, and works still more. Night after night, the infected bee lives in exile, yet still toils for its family until—in an ideal world—it perishes without the fly larva having fully developed within its belly, pupated in its corpse, and gone on to infect any number of other bumblebees.*

Why, though? Why would the bee go through the trouble? Well, because of a little something called the meaning of life, which I have managed to figure out and will now share with you: Pass your genes to the next generation at any cost. For the infected bumblebee, it's more about helping ensure the survival of her queen mother, the only one in the colony who can breed. But throughout the animal kingdom, life is sex—the dogged, more-often-than-not bizarre quest to reproduce. Which brings me to the slightly modified meaning of life for our zombifiers, which I have also figured out: Pass your genes to the next generation at any cost, and if that means destroying the sex life of your host, then so be it.

---

*Such struggles are not unlike those of Sheila Hammond in the show *Santa Clarita Diet,* the diet being humans since Sheila is a zombie. She's what I call a cognizant zombie, who, like the bumblebee, is aware of her infection and embarks on a quest to extend her life as a rotting monster. In the bumblebee's case this is for the good of the colony and in Sheila's case for the good of her family. This means Sheila has to deal with challenges like her toe falling off, but unlike the zombie bumblebee, at least she still gets to sleep at home. Count your blessings, I suppose.

6

# Dawn of the Sexually Undead

***Let's talk about sex, baby, let's talk about the parasites that steal the most precious thing you could ever steal from an animal: its ability to reproduce.***

Males and females have their differences. I'm not talking about the *Men Are from Mars, Women Are from Venus* kind of nonsense—relationship books and their stereotypes are oh-so-human constructs. It's just that, as is the case for a lot of species, the Y chromosome haves and have-nots of the human world can sometimes look dramatically distinct. This is known in biology circles as sexual dimorphism.

But we've got nothing on one variety of zombifying insect that has over evolutionary time taken this contrast to the extreme, developing into males and females that look so different, they may as well be different species. These are the strepsipterans, flylike parasites that make life miserable for social insects like bees and wasps and ants by invading their bodies, manipulating their minds, sterilizing them, and spewing larvae out of their abdomens.

A male strepsipteran is a beautiful thing, with great big eyes and wings that look like rippling fabric (*strepsiptera* meaning "twisted wing"). The female, though, doesn't screw around with flight—or eyes or mouthparts or legs or antennae. She's a bag of eggs, looking more like a flattened worm than anything. And that suits her fine: The only home she'll ever know is the belly of another bug.

For our purposes, let's focus on the home that is the paper wasp, and the strepsipteran *Xenos vesparum*. The female doesn't just grow in the wasp's abdomen like a nematomorph does in a cricket. No, she also grows *through* the exoskeleton, poking her rump out of the host's layers of armor. When she's ready to mate, she releases a pheromone that sucks in males, which find her exposed oviduct and do their deed. Having completed their sole purpose in life, the males perish. They'll have lived only a few hours, never feeding or sleeping or doing anything at all besides furthering their genetic line.

You might expect, then, that natural selection would favor a strategy that gets males and females as near to each other as possible so the search for mates is less of a hassle. And in fact something peculiar happens to paper wasps harboring strepsipterans. In the summer, infected worker wasps—all females, like with ant colonies—desert their nest and congregate on vegetation out in the open. (Not that they were much help to the colony to begin with: Infected workers tend to laze around. That's probably good for the parasite from an energetics perspective, as laziness means more energy for the parasite to exploit.) And not at random, either. They form clusters in the same spots where paper wasp males usually gather to catch the eye of females.

What's important to consider here is that, unlike a lot of social insects, paper wasp worker females can reproduce. Think of it like an insurance policy for the colony—if the reigning queen should die, the next in command takes her place as the mating female. But fertility is in its own way a burden, considering how much energy it takes to develop eggs. Same, too, for fertile males, though in fairness producing sperm is less of a burden than producing eggs. Still, if your host is no longer trying to reproduce, you can co-opt that saved energy for your own ends. So if a strepsipteran lands in a female wasp, it'll sterilize her. If it lands in a rarer male wasp, it'll castrate him, too. That might seem to work against the parasite, since theoretically the host could use that saved energy to fight off the invader, but there will be no bouncing back from this.

What you end up with is a mass of wasps that may appear to be gathering to mate, but are in reality all sterile. The parasite may thus be corrupting the insects' reproductive instincts, driving the hosts out of the colony to these traditional mating grounds at the wrong time of year. The male parasites erupt from the wasp hosts, killing them, then find another wasp nearby with a female strepsipteran inside to inseminate. Those wasps will live on, carrying the fertilized parasites within them back to the colony. Here, the strepsipteran releases her larvae—they look like pill bugs, only with longer legs—which drill into the wasps' larvae and grow fat, surviving as the grubs metamorphose into adults. Thus the cycle begins anew.

Zombifying a wasp to drive you to an orgy is downright devious, but I want to return to the sterilization bit, because it's a not uncommon manipulation among parasites. This can come about either by the parasite consuming its host's sex organs, by the parasite releasing some kind of chemical, or by the burden of a parasite simply wearing down the victim's body. The idea being that the host can temporarily sacrifice reproduction to save energy, in the hope that whatever is ravaging its body will stop with the ravaging. And do consider that sex isn't only about expending energy producing eggs or sperm. An organism also has to spend time searching for a mate . . . and males often have to put on displays . . . and even growing things like ornamental feathers to impress mates takes resources. Forgo all that and the host might be able to mount a solid defense against the parasite, or co-opt the saved energy to ride out the intruder. Problem is, that's not going to work against the strepsipterans, which have coevolved with the wasps to assume command of their minds and dodge their defenses.

Which brings us to the meaning of life. If anyone ever tells you that it's to just be kind to people, tell them to get bent. Because evolutionarily speaking, the meaning of life on Earth has and always will be to pass your genes to the next generation—that is, reproduce as much as you can. Kindness is a human construct. I mean, by all means, be nice to each other, but don't think for a moment that you're on this planet for any reason other than to make babies. Reproduction is—how shall I say this—*a matter of some consequence*. So what's fascinating about a sterilized zombie is that from an evolutionary perspective, it's functionally dead. It will never reproduce, and if it can't do that, there's no point existing. The host's genes may have originally coded for its body, but if it didn't manage to reproduce before it got infected, its genes are lost forever. The zombie becomes a mere costume for the only genes still in play—those of the parasite. It's the sexually undead.*

The ability to mate, then, is the greatest thing you could ever steal from an animal. And you better believe that our zombifiers have figured out how to do it in fantastical ways.

## Barnacles: The Sexual Deviants of the Sea

If it weren't for Mr. Arthrobalanus, the biological sciences as we know them might not exist. Charles Darwin might not have had the courage to publish his society-shaking theory of evolution by natural selection. And it might have been that Darwin didn't learn so much about barnacle penises.

Darwin met Mr. Arthrobalanus in Chile in 1835, during his famous voyage on the *Beagle*. And by "met" I mean collected, for

*Sex is not a common theme among the zombies of pop culture, probably for two reasons. One, gross. And two, the zombie virus spreads by bites, so reproduction among the hosts is moot anyway. The virus's time is better spent getting the zombie to gnaw on the uninfected. For the real-life sexually undead, though, things are rather more complicated, as we shall see.

Mr. Arthrobalanus was Darwin's pet name for a barnacle he gathered on a shoreline in the Chonos Archipelago. He found the orange little thing not stuck to a rock like you might expect of a barnacle, but drilled into the shell of a Chilean abalone, actually a kind of snail. Strange goings-on for a barnacle, which Darwin called both a "singular little fellow" and an "illformed little monster" (at least he was consistent about its size).

Darwin returned to England and of course formulated his famous theory—and promptly stowed the manuscript in a drawer. He knew what his grand idea would do to science and society, and he agonized over it. Not helping matters was the fact that Darwin's friend the botanist Joseph Hooker suggested in a letter that to theorize about species means getting to *know* species. Replied Darwin, "How painfully (to me) true is your remark that no one has hardly a right to examine the question of species who has not minutely described many." Luckily, though, Darwin's other friend—Mr. Arthrobalanus—appears to have been the kick in the pants he needed. The soon-to-be father of evolutionary theory set out on an epic quest to catalog barnacles (which are crustaceans, not mollusks like clams and mussels), dedicating eight years of his life to teasing the things apart under a microscope in his study. It became such an obsession that Darwin's young son George, upon inspecting a neighbor's home, reportedly inquired: "But where does he do his barnacles?" (George was a weird kid.)

As it would turn out, Mr. Arthrobalanus wasn't a mister, but a female. And tucked inside Mr. Arthrobalanus were teeny-tiny males, mere bags of sperm with little other function—like the female strepsipterans, they had no mouth or intestines. What each male did have, though, was "an immensely elongated prosbosciformed [read: reminiscent of an elephant's tusk] penis, coiled up and filling the rest of the inside of the sack down to the testis, which latter occupies the whole anterior, and generally lower end of the animal." Meaning, the male barnacle consisted almost entirely of genitals, an accomplishment not lost on Darwin. "I should think that this organ could be extended by the animal to, perhaps, even the 100/1,000th of an inch—that is, to between eight and nine times its own entire length!" And not just to show off: With such a prodigious penis, the minuscule male could stretch to reach the female's individual eggs, which about matched him in size. (Your "traditional" male barnacle on the shoreline will do the same, extending his penis to fertilize his neighbors. But nothing in the animal kingdom beats Mr. Arthrobalanus in terms of relative penis size.)

With Mr. Arthrobalanus playing the part of muse, Darwin published an exhaustive four-volume study of the barnacle family. And perhaps it gave him the extra bit of mettle to publish *On the Origin of Species* once Alfred Russel Wallace sent that fateful letter threatening (accidentally) to scoop him on his theory. What the barnacle survey *did* do—we can say with confidence—was kick sand in the face of Victorian ideals. You know, men powerful, women submissive—in no way giant females sexually imprisoning tiny males. Barnacles, Darwin found, are universally bizarre with their sex, a pursuit that he knew was central to his theory. Without sex, you couldn't have the extreme diversity of species on Earth. You can reproduce asexually, of course—all manner of plants and animals do it to clone themselves. But what's nice about sexual reproduction is the variation we talked about in the first chapter. Two parents combining their genes leads to more variation, since the code combines differently for each offspring.

If the Victorians had only known the half of weird barnacle sex, it would have blown their minds, because a particular variety takes the whole circus and throws in a healthy dose of body snatching, sterilization, and sex changes. These are the rhizocephalans—barnacles that have turned on their own crustacean kind. While the males of Darwin's beloved barnacle species are exceedingly tiny and shell-free, the same goes for male *and* female rhizocephalans. That is, until the female assumes the body and mind of the crab she invades.

She begins as a microscopic oval of a larva, like all barnacles, floating around with other plankton in the ocean. Whereas your typical barnacle needs to settle on a sturdy surface to develop into an adult—they'll take anything they can get, really, including marine giants like whales—the rhizocephalan is burdened with finding a crab host in the great big sea. Should the female be so lucky, she still has to find a chink in the crab's armor, so she lands and seeks out one of the many sensory hairs that cover the body, then anchors herself with cement.

That is where her similarities with other barnacles end, because she next stabs into the base of the hair with a needle-like appendage and injects something called the vermigon, aka the primordial parasite—essentially cells of herself. The rhizocephalan then starts growing throughout the crab's tissues as a kind of root system, not unlike how *Ophio* sets itself up in ants. And she grows and grows, snaking through the entire crab—the abdomen and organs and claws and legs—all the while soaking up nutrients to fuel her proliferation. Eventually, this all becomes too much for the host to handle: Typically it sheds its exoskeleton every so often to make room to grow, but that stops. Which is exactly what the barnacle wants, for she needs to now poke a bit of herself out of the crab, and she can't do that if the shell keeps popping off. She forms a bulbous, yellow-orange mass at the rear of the host's belly, where a female crab would normally hold her own eggs. (Think of this as a reverse shrub, which is nourished by roots growing through the crab's tissues as opposed to soaking up water.)

Maybe it's the barnacle messing with the crab's brain, or maybe it's the crab suffering some sort of automatic physiological response, but the host starts grooming the parasitic sac like it's her own offspring. In fact, if you prevent a host from tending the sac, the parasite will rot away. And if the rhizocephalan happens to invade a male crab, she pulls an extra trick. Male crabs are smaller than females, but when the barnacle gets into a male, she'll feminize him, transforming his testes into ovaries. She also triggers the abdomen to widen to match the size of a female crab, probably because this affords the parasite more room to grow without spilling over the edge of the exoskeleton. Then the male, too, will groom the sac as if he were a female tending eggs, which is particularly puzzling because it isn't as if the barnacle is tapping into some sort of instinct for the male—he hasn't a clue how to look after eggs. But be the crab a male or a female, one thing is for certain: The rhizocephalan sterilizes her host, freeing up more energy and in effect replacing it in the ecosystem as a reproductive being. She proceeds to order her host into deeper waters, where the crab won't have to compete with its uninfected comrades for food, thus helping it get enough energy to sustain the parasite within.* And it's here where the rhizocephalan will find a mate.

---

*At least one rhizocephalan, though, *suppresses* the appetite of its host, the flat-backed mud crab. Healthy crabs eat eight times the number of mussels that infected crabs do. And considering that almost a quarter of the mud crabs in a given environment harbor rhizocephalans, the parasite makes a serious impact on its food web. So in case I needed to say it again: Manipulative parasites can alter their ecosystems in dramatic ways that

The male rhizocephalan also starts out as a minuscule larva, but his task is doubly difficult as the female's. He has to find not only a crab, but a crab that's already harboring a female rhizocephalan, and a virgin female at that. Should he succeed, he'll enter the sac, but he doesn't bother joining her in growing as a root system. Instead he transforms into a mere testicle and imbeds himself inside her because, like with Darwin's barnacle, males need only concern themselves with mating. Thus nestled in unholy matrimony, the barnacles bring their eggs and sperm together. (As such, they are now essentially a self-fertilizing hermaphrodite, and will live the rest of their lives bonded as a single organism, which itself has bonded with the host to create an animal that is actually a trifecta of beings.) Inside the sac, the fertilized eggs hatch into larvae, which the crab boosts into the world by shaking its rump, just as a crab would to disperse its own eggs.

How, though, can a barnacle induce the sexual transformation of a crab? The answer appears to be twofold: with brute force and chemical warfare. The rhizocephalan is growing so extensively through its host not just to soak up more nutrients, but to infiltrate the organs and central nervous system. The feminization of the male crab, for instance, is likely the result of the parasite monkeying with the androgen gland, which is responsible for sex differentiation. And destruction of a female crab's ovaries is probably part chemical sterilization and part physical trauma. But experiments show the many effects of the rhizocephalans are likely more chemical than anything: Inject extracts from the barnacle's roots or the blood of an infected crab into a healthy

---

scientists are only beginning to understand. But that's all great news if you're an uninfected crab, because it means less competition for food. Bonus: You don't have to suffer a barnacle growing through your flesh.

crab, and it'll begin transforming as well. Exactly what chemicals are in play here, though, remains unknown.

What's clear is that the rhizocephalan-infected crab is, more than any other victim of a manipulative parasite, not just a vehicle but a *sexualized* vehicle. The barnacle snatches away its host's ability to reproduce, then uses the crab to further its own sexual means. The irony is that all the while the crab is acting as if it's reproducing, when in fact as a sterilized animal, there's no point in it existing any longer—if it wasn't able to mate before infection, it's failed its one mission in life. Yet even after the crab is done helping bring baby parasites into the world, it'll live on despite its various traumas, roaming the seafloor as the sexually undead.

## If a Bacterial Infection Changes Your Sex, It's Probably Too Late for Antibiotics

A rhizocephalan may begin her time on Earth as a mere life-speck in hundreds of millions of cubic miles of ocean water, but she grows to fantastic proportions—at least as far as her poor host is concerned. Other parasitic manipulators, though, operate on a much, much smaller scale. You see, even bacteria can alter their hosts' sex lives. And no bacterium does this more bizarrely—or bizarrely commonly—than *Wolbachia*, which may infect up to two-thirds of all insect species, not to mention a multitude of other arthropods like the arachnids, making it perhaps the most successful parasite on Earth. And by that token, it's by far the most successful manipulator.

Whereas a rhizocephalan hacks the reproductive instincts of its victims, the various strains of *Wolbachia* transform the very nature of their hosts' sex lives. *Wolbachia* doesn't so much transmit from one insect rubbing up against another, instead passing from mother host to daughter and then to the next daughter, on down the line. In this way the bacterium hijacks the most powerful force on Earth—sex—to spread itself around. Specifically, it stows away in eggs, surviving as those single cells divide and diversify into offspring made of countless cells—muscle cells and brain cells and exoskeleton cells. Yet even with all that chaos, *Wolbachia* always ends up lodged in the eggs of the daughters. So the parasite lives intertwined with the host, be it a fly or beetle or spider or what have you, over the generations, altering the fate of a distantly related species not only on an individual level, but over lineages stretching millions of years.

Here's the problem, though. *Wolbachia* gets from mother to daughter fine by hitching a ride in an egg that transforms into a daughter with her own eggs to exploit. But males are trickier: The bacterium can't cram itself into sperm cells because they're too small compared to an egg. So it would seem that for *Wolbachia,* male hosts are at best worthless, and at worst a dead end. But that's not giving the parasite any credit, because *Wolbachia* disrupts its hosts' sex lives in four unique and cunning ways, depending on the strain and what species it infects.

First, and most simply, the bacterium just kills all the males during embryonic development. For instance, in fruit flies the bacterium sabotages the male embryo's very genetic structure, known as chromatin, which is responsible for compacting extremely lengthy strands of DNA into something that fits inside a cell. Thus *Wolbachia* corrupts the instructions for building a healthy fly, assassinating the male before he has much of a chance to develop. The message is clear: Don't you dare compete with your sisters. So young infected female flies will emerge into a world where they need not fight with brothers over food, making them all the more likely to survive and breed and ferry the bacteria to the next generation.

*Wolbachia*'s second strategy for getting rid of males is known rather poetically as parthenogenesis, meaning "virgin creation" (and also known rather less poetically as thelytoky—from the Greek, meaning "female offspring"—so let's stick with parthenogenesis). Infected females not only produce solely female offspring, but do so without fertilization. Infected mites, for instance, simply lay eggs that hatch into girls identical to themselves, no males required.

*Wolbachia*'s third strategy doesn't eliminate males, but instead exploits them. In this case the bacterium modifies—though, again, does not take up residence in—their sperm. If an infected male mates with an uninfected female, these modified sperm destroy her eggs instead of fertilizing them. If a male infected with one strain of *Wolbachia* mates with a female infected with a different strain, the same thing happens. But if a pair of insects infected with the same strain shack up, their eggs and sperm fit together like a lock and key: The bacteria in the eggs defuse the killer sperm, leading to viable offspring. The young females among them will carry the *Wolbachia* in their own eggs, and the males will fly around with that special key. So instead of outright killing the males, this strategy puts them to work for the correct strain of *Wolbachia*. Because only females can pass the bacteria to the next generation, if a female is harboring the wrong strain, a male that fertilizes her is helping propagate the competition. A big no-no, as far as the parasite is concerned.

The final strategy of *Wolbachia* also exploits the males instead of killing them—by turning them into females, like with rhizocephalan-infected crabs. More remarkable still, it seems to trigger this the same way the barnacle does to the crab, by targeting the androgen gland—at least, that is, when *Wolbachia* gets into pill bugs or roly-polies or whatever you call them. This makes sense, as pill bugs are terrestrial crustaceans, and therefore cousins to the rhizocephalans' crabby victims. But the bacterium mucks with the androgen gland—which, again, is responsible for sex differentiation—in early embryonic development, much earlier than in the case of the barnacles. So what would have been a male pill bug instead develops into a female capable of carrying the *Wolbachia* to the next generation.

Now, all of these strategies involve *Wolbachia* physically manipulating its hosts into the sexually undead. But there's also evidence to suggest that the bacterium is messing with its victims' minds as well, in particular among fruit flies. One group of researchers found that if you cure flies infected with the key-doesn't-fit-the-lock variety of *Wolbachia*, their mate discrimination drops by 50 percent. Meaning, infected flies are choosier about who they partner with, which could be a manipulation to get the bacterium into the right kind of infected mate. Another study determined that male fruit flies infected with the same variety of *Wolbachia* mate at a higher rate. If the supercharged male gets together with an infected female, great, the key fits the lock. If he mates with an uninfected female, also great for the *Wolbachia*, because his sperm kills her offspring, meaning less competition for flies infected with the bacterium.

Given how widespread *Wolbachia* is—up to two-thirds of insects, remember—and given how zombification has so readily evolved across the animal kingdom, I find it hard to believe that the bacterium's behavioral manipulations are limited to fruit flies. In fact, it's an impossibility, now that I think about it. *Wolbachia*'s job is to spread itself at any cost, which it does with one of four clever techniques for physically manipulating its host's sex life.

But why would it stop there? As we've seen with the wasps and fungi and worms, messing with the brain gives the parasite a reproductive advantage over its peers. It's why an acanthocephalan can manage such a complex life cycle, and why *Ophio* can better position its spores to rain down on the ant colony. I suspect that elsewhere among insects *Wolbachia* is up to zombie shenanigans that are far more elaborate and bizarre than monkeying with the sex drive of male fruit flies. I mean, no disrespect to the researchers here, but studying a fruit fly is relatively easy—it's a model organism that scientists keep in labs all over the world. *Wolbachia*'s exploits outside the lab, out in the insect realm, must be creative indeed, given all the different ways bugs reproduce. Think about how strange the habits of just one group of animals, our strepsipterans, are. Now consider that strepsipterans, too, can carry the bacterium. Might the zombifier fall victim to a zombifier itself? How might the bacterium go about manipulating the strepsipteran sex life to its own ends, should new conditions make it necessary to do so? It would certainly take inceptulation to a whole new level.

To recap: *Wolbachia* hates males. And as well it should, for males stand in the way of the bacterium propagating itself, all because the bacterium can't fit into sperm. So depending on the species it torments, *Wolbachia* has evolved crafty ways of either dispatching or transforming males into something else entirely—the sexually undead. Just as your traditional Hollywood zombie exists solely to spread the virus or bacterium that's taken command of its body—the hyperaggression, the biting, the refusal to die—a *Wolbachia*-infected male is biologically reprogrammed to maximize the parasite's odds of furthering its genetic line. And more than any other parasite we've met so far, *Wolbachia* has the potential to go full-tilt pandemic on insects.

Which is one zombie outbreak the human race might actually

appreciate, for scientists have figured out how to weaponize *Wolbachia* to fight the deadliest animal on Earth: the mosquito. The diseases this pest transmits, such as malaria and dengue fever, kill upward of a million people a year around the globe. But like the majority of insects out there, the mosquito has its own enemy in *Wolbachia*. The problem is, a particularly invasive mosquito that's menacing the United States, *Aedes aegypti*, is one of the minority of bugs that doesn't naturally carry the bacterium.

So a biotech start-up called MosquitoMate has convinced it to. Researchers pulled the yolk out of an egg from an *Aedes albopictus* mosquito, which carries *Wolbachia*, and injected it into an egg from an *Aedes aegypti* mosquito. That egg then hatched into an "Eve" mosquito that was able to pass the bacterium on to her children, which passed it on to their own children, and so on. So what MosquitoMate has now is a sustainable population of Judas bugs—and all it took was initially engineering one individual.

MosquitoMate ships its creations to local mosquito abatement programs, which each release hundreds of thousands of the infected males over a few summer months (only female mosquitoes can bite, so no worries there). The idea is that infected males will outcompete uninfected males in the area for the affection of *Aedes aegypti* females. These females of course don't naturally carry *Wolbachia* because no one has engineered them to, so if they mate with infected males, the bacterium works its usual magic, sabotaging the offspring. By thus releasing a horde of the sexually undead, MosquitoMate claims it can crash mosquito populations by as much as 80 percent. It's not total eradication, sure, but any zombie apocalypse has its share of survivors.

## Sex Appeal

And so the sexually undead roam the Earth. They join the larder zombies like the cockroaches, which live to be eaten alive. And the bodyguards, which die while protecting the parasites that have erupted from their bodies. And the full-tilt puppets like the zombie ants and vehicles like the worm-laden amphipods and crickets. A dizzying range of strategies for a dizzying array of life-styles. On our journey through the world of the zombifiers, though, it's the sexual manipulators that force us to confront the true meaning of life on this planet.

I don't have kids, but I suppose I get it. They're cute, at some point they become useful, and if you don't screw things up too bad they might take care of you when you're old. All interesting ways to justify what is the simple meaning of life: You are programmed to reproduce. You're programmed to make yourself presentable to potential mates, you're programmed to eat so you have the energy to procreate, you're programmed to respond to your baby's screams because that makes it harder to leave the kid sitting on a rock somewhere. And it's not just you. The whole of the animal kingdom exists to propagate itself. It's what life does. So for our zombifiers, sex is an immensely powerful weapon because sex is an immensely powerful force.

Really, sex is what drives the zombifiers to do what they do in the first place. The blob of a female strepsipteran and the branching roots of a female rhizocephalan barnacle have over evolutionary time sacrificed what we might consider a "normal" body so they'd have a better chance of passing down their genes at the expense of the host they're embedded in. Sex is what did this to them. The strepsipteran doesn't need to fly and the barnacle doesn't need to cement herself to a surface because they're exactly where they need to be—inside a host that nourishes them so they can develop their young. Sterilizing the host is an extra indignity, but I'll be damned if they aren't better mothers for it.

But the strepsipterans can only make this happen if males and females are able to find one another, in this case by smell. The two parties share a way of experiencing the world around them, choosing one stimulus among many exuding from this chaotic planet of ours. Other creatures skip smell altogether, instead preferring sight or sound or touch. Within all these senses, each species is tuned to a specific range—say, ultraviolet or visible light, or different sound frequencies—so every animal has a highly subjective way of making sense of an objective world. And you can be damn sure the zombifiers have figured out how to exploit that.

# 7

# The Great Escape from the Umwelt

***You think you know how this world of ours looks and feels and smells? The body snatchers would like you to know that you know nothing.***

You, my friend, are a tick. That'd make you an arachnid, though, unlike your spidery comrades, you sport six legs, not eight—at least for the time being. Instead of bones for support, you have an exoskeleton, which serves as armor against predators and pathogens alike. With very little head to speak of, you're pretty much all teardrop-shaped abdomen. And that suits you fine.

You've just emerged from an egg as a larva during what human aesthetes call *the magic hour*, when the setting sun colors the woodlands gold. Not that you can appreciate it, considering you're nearly blind and cognitively ill-equipped to appreciate anything other than blood. But no matter. You're famished and on the prowl for a host—something like a bird or mammal will do nicely. So you make your way from the dirt where you were born up a blade of grass and wait. And wait and wait. A week, a month, maybe more. You're questing, as humans call it. And you have the patience of a tick, my friend.

There, a vibration. And there, the scent of butyric acid, the essence of a mammal that in large quantities a human might describe as resembling sour milk or farts. A sensor array of hairlike structures in your front legs, known as Haller's organ, has helped you sniff out signs of life. The scent and the vibrations grow stronger, and stronger still. *Rumble rumble, sniff sniff,* until, at last, the setting sun hits your mark and casts a shadow. It's a coyote, not that you know it as such. And fatefully, it ambles past, rubbing against the blade of grass you call home. You grab its fur and hold on as if your life depends on it—because it does: You must feed only once to reach the next phase of your life.

You sense the coyote's heat. You sense the humidity of the air trapped in its fur. You sniff out the carbon dioxide it exhales. Clearly, you're in the right place, so you burrow through the coyote's hair and slice into its flesh. If you like what you taste, you settle in to feed. If not, you might wander elsewhere on the coyote and try again. You drink and drink as the days go by, your abdomen expanding until it can expand no more.

You drop off your gracious if unwitting host and return to the earth, transforming into a nymph and growing that extra pair of legs. Finally, you're a real arachnid. You climb up into a bush and quest once more. You sniff, you look for shadows, and—if you're lucky—you find another host like a deer and complete the next phase of your life. Back into the earth to become an adult, then back onto a host for your final meal. *Sniff, feel, taste . . . sniff, feel, taste.* This is your existence.

It's a world that couldn't be more different from our own. You and me (you can stop pretending to be a tick now), we're visual animals. Our world is more than the mere shadows of the tick's, one of colors and contrasts and, thankfully, magic hours. Our eyes and brains can distinguish millions of different colors. Human vision is so fantastic, in fact, that at night we can see candles thirty miles away. And while we have nowhere near the night vision of a nocturnal creature like a raccoon, our eyes are still remarkably capable at gathering what scant evening light is available.

Humans and ticks live in the same world, yet for the two species, that world is two different existences. For a human, it's a life of light and sound, with a complement of smell. For the tick, life is full-tilt smell—specifically the smell of butyric acid—combined with vibrations and to a lesser degree shadows. So while Earth is one objective world, with trees and rocks and rain and earthquakes and oceans, within that one objective world are millions upon millions of subjective worlds. Every species lives in an existence that another does not, and even within those species, individuals vary in their abilities. Your umwelt is subtly different from my own because you might hear or see better or worse than me. No two organisms experience reality in exactly same way, and that has big implications for how we contemplate the world of the mind snatchers. Not only can we barely fathom how the zombifiers manage to survive within their hosts, but we have little clue as to how they perceive their world. And that bothers me.

## The Legendary Biologist Who Thought Darwin Was Full of Shit

It'd be fair to say that few people before Jakob von Uexküll cared about how a tick experienced the world. Uexküll was born in Estonia in 1864, into a long line of German Baltic nobility. He became a child of nature, spending time with caterpillars and frogs and such—standard fare. At the age of twenty he enrolled in college for zoology and, like so many of us, made the mistake of reading Immanuel Kant. Thus a peculiar sort of chimera formed: a biologist with the proclivities of a philosopher, right down to growing the wild, Nietzsche-esque mustache. And indeed, Uexküll both infuriated the field of biology and contributed an idea that's become central to the study of animal behavior, particularly our subfield of zombification.

For Uexküll, the tick was the embodiment of what he called the *Umwelt* (plural: *Umwelten*). Well, I suppose a lot of German speakers say *Umwelt* fairly often, considering it means "environment." But Uexküll's revelation was that each organism in a given environment subjectively constructs that environment. "Just as a gourmet picks only the raisins out of the cake," he wrote in his 1934 work *A Foray into the Worlds of Animals and Humans,* "the tick only distinguishes butyric acid from among the things in its surroundings." The tick, just by existing, creates its own umwelt, distinct from the umwelt of the coyote or deer or human it draws blood from. It doesn't mind that you and I can see so well. Its existence is smell and vibrations and heat, and it sits comfortably in that particular umwelt. By that logic, we humans don't mind how the tick experiences the world—well, save for Jakob von Uexküll.

So, Uexküll argued, in a given ecosystem there is no objective reality, just the multitudinous umwelten of its inhabitants. "It seems, then," he wrote, "that we must abandon our fond belief in an absolute, material world, with its eternal natural laws, and admit that it is the laws of our subject which make and maintain the world of human beings." Meaning, an objective world is a fallacy, which from a biological perspective is both brilliant and totally wrong. The tick does its thing and we do ours, occupying distinct umwelten based on the senses available given our biology. And sure, we build different worlds as species, true enough. But to write off an objective reality is all too human: This existence isn't made for us, or any other creature for that matter. We are no more consequential than the tick.

So it's no surprise that Uexküll wasn't exactly fond of Darwinian theory. How not fond, exactly? Well, he once wrote what could be the most ironic sentence in the history of biology: "Darwinism, the logical consistency of which leaves as much to be desired as does the accuracy of the facts on which it is based, is a religion rather than a science." That'd be the same Darwinism that—with regard to genesis, at least—made religion somewhat negotiable by showing how species could come to be without a creator. But what Uexküll was so pissy about was the inherent lack of a plan or purpose in natural selection, the idea that species evolve largely due to randomness and without the benevolence of God with a capital "G" in the Christian sense. And this was in the early twentieth century, mind you, when Darwin's theory of evolution by natural selection had taken its hold in science.

But I'm being negative, and we're here to talk about the good stuff. So I'll say this: We cannot fully understand and appreciate the world of the body snatchers without Uexküll's concept of the umwelt.

Consider the amphipod and its acanthocephalan parasite, and consider the senses at play here. The scene is a pond of moderate depth and murkyish water, the bottom being dark and the surface being relatively clear. The amphipod's got peepers, so it senses it needs to stick to the dark depths to avoid ending up in the stomach of a bird, which itself has eyes it uses to hunt. The amphipod also has its antennae, which pick up chemical cues and vibrations in the water, further helping the creature avoid predators. The acanthocephalan curled up inside the amphipod, however, cannot see and it cannot feel. Its world is darkness.

You know how this story ends. The parasite drives its host out of the depths and into the shallows, where the keen-sighted bird strikes. It's the confluence of three lives and three ways of experiencing the world—three umwelten colliding. In a single objective reality, a trio of subjective existences meet. And at the very center, the zombifier has hijacked the senses of its host and exploited the senses of its host's enemy.

This is the essence of mind snatching in the animal kingdom: Controlling your zombie isn't about turning it into a simple brain-dead automaton. It's about assuming its senses, its umwelt. The acanthocephalan cannot see, and really, it doesn't need to. It commandeers not only the amphipod's brain, but also its sight, guiding the victim into the belly of a bird. Thus a host's senses, which are indispensable tools for keeping an animal alive, are also a critical weakness, because senses, from sight to smell, are ripe for exploitation. Body snatching is so common and so creative in the animal kingdom in part because there are so many different dupes with so many ways of perceiving the world—and therefore so many weaknesses for the zombifiers to exploit. Which means zombification must be far more common than we realize.

I mean, Plato pretty much said so.

## The Actuality of the Cave

Plato—ever the optimist—had this idea of prisoners kept in a cave from birth, chained to a wall by their feet and necks so they can only look straight ahead at another wall. Behind them burns a fire, and between them and the fire is a walkway, where puppeteers perform, casting shadows on the wall for the inmates to see. All the poor prisoners will ever know, their entire reality from birth to death, are shadows and echoey sounds.

Until, that is, someone drags one of the captives out into the real world. The man's eyes take some time to adjust, but when they do, a convoluted world emerges. Dazzled, the erstwhile prisoner returns to the cave, where he is once again blinded in the darkness, and says to his erstwhile comrades, "Did you know there's an entirely different reality up there, with trees and rocks and stuff and also a thing called the sun?"* Certainly they would ridicule the freeman, and certainly fear this world, for it would

*This would have been in Ancient Greek, of course, not English.

make them blind and unable to see the shadows in their subterranean home, the only reality they've ever known. Plato's whole point being, in his allegory of the cave, that our subjective perception of the world may not be all there is to knowledge. The true philosopher transcends the human "reality."

But let's move from the allegory of a cave to what we might call the actuality of the cave, namely a real cave devoid of prisoners. Here live the troglobites (not necessarily zombifiers, but let's broaden our horizons for a moment), animals that, like Plato's prisoners, never see light. These include blind fish, natural selection having over the millennia whittled down their peepers. (Again, there's no such thing as "progress" in evolution.) So instead of sight, these fish rely on a finely tuned system running along their bodies, known as a lateral line, to pick up the minute changes in water pressure their prey produce. Nearby, the cave salamander goes about things a bit differently, instead sensing the electrical signals of its prey, as a shark would. And crawling above the fish is the cave cricket, which feels around with super-elongated antennae.

Unlike the prisoners in the allegory of the cave, the creatures in the actuality of the cave have no chance of ever seeing the light of day, the "real" world that humanity inhabits. In fact, this world would kill them pretty much immediately, just as their world would dispatch us. Their reality is darkness—permanent midnight. They feel their way around and sense electricity and pressure without ever glimpsing a photon. In the actuality of the cave, the troglobites' darkness is as incomprehensible to us as our world of light is to them.

Yet even up here in the world of light, scientists stumble around in pursuit of the zombifiers. We're visual creatures, after all. A scientist's default observation of the natural world is sight, with the supplementary senses of smell and hearing and touch. I can watch a nematomorph erupt from the abdomen of a cricket, easy enough—a parasite and its host have wandered into my umwelt. But the parasite and its host are also sensing on levels I'll never understand. I can't sniff out a river to leap into, for instance. And I certainly don't know what it's like living inside a cricket's abdomen, though I think it's safe to say it's dark and wet. We humans are too wrapped up in our own world, too biased not just toward light, but toward a tiny portion of the electromagnetic spectrum of visible light.

We're ignorant to the vast majority of environmental cues—even aboveground we're stuck in a cave. Which is fine, to a point, because our heads would explode if we tried to process all the possible input in the world. But it means we're ignorant of the vast majority of manipulations among parasites. "The natural world may abound with parasite-induced changes that are not conspicuous to the human senses, changes that scientists do not even imagine, much less seek," writes the biologist Janice Moore in her foundational work *Parasites and the Behavior of Animals*. To find them, science has to transcend the darkness. And a good place to start is the world of smell, specifically the bird with a nagging case of BO.

## The Curious Case of the Bird That Smelled Kind of Funny

The red grouse leads a precarious life. It's a beautiful little thing—colored all auburn with bright red eyebrows—that calls the moorlands of Britain home. And indeed the humans with whom it shares a country call it the moorbird, which, let's be honest, is a much better name so I'll go ahead and use that from here on out. The moorbird shuns trees, and just as well, for trees are in short supply here anyway, so the creature risks nestling into the grass, where plenty of mammalian predators like foxes roam. Critically for our story of moorland parasitism, included among those predators are human hunters and their dogs.

Mammalian predators, though, are really the unwitting henchmen to the moorbird's archenemy: the nematode *Tricho-strongylus tenuis*, aka the strongyle worm, which, let's be honest, is a much better name, so I'll go ahead and use that from here on out. This little nasty has a direct life cycle, requiring the moorbird—and the moorbird only—to complete its development from egg to reproductive adult. The strongyle lodges itself in the bird's cecum, a sort of pouch in the intestine, where it mates with its compatriots and releases eggs, which ride with the feces out of the digestive system. (In case you were curious, you can tell a strongyle-infected dropping from a regular one if it's sticky, as opposed to being more fibrous. And should you be so bold as to investigate, you can tell a strongyle-infected moorbird from its swollen cecum.) The eggs hatch into larvae right there in the feces on the moor, then crawl up onto the tips of heather plants—shrubs that bloom a brilliant purple. And heather plants, as it so happens, are the moorbird's favorite food. Larvae consumed, life cycle complete.

Hunting moorbirds has been all the rage in England for centuries, and a moorbird hunter's greatest tool has all the while been the hound, which, in accordance with its umwelt, is far better at sniffing out birds hidden in the grass than humans are at spotting them. So in the early 1980s, three scientists set out to test their theory that mammalian predators would be better able to track down strongyle-infected birds than uninfected ones. They began by scouring the plains of Gunnerside Moor, near Leeds, for dead birds. To determine cause of death, they were working with a straightforward binary: moorbirds that had either been mauled or had died from something else. If you've never come across the body of a bird that's been mauled to death, signs include missing feathers and flesh. So the scientists looked for that. Other birds they unfeathered and scoured for bite marks. They took all these back to the lab and washed out the cecum of each bird—at least, each bird that was intact enough—over gauze to catch and count the strongyle worms. Intriguingly, birds that had fallen victim to predators carried far more worms than those that had passed away from other causes.

In the spring of 1983 and 1984 the scientists scoured the moor at night for live birds, carrying powerful quartz halogen lamps to dazzle their targets and capture them, like deer caught in headlights. Back at the lab, they treated some of the moorbirds with a dewormer and another, control group with only water, outfitted them all with radio trackers, and returned them to the wild. Then the scientists, with the help of two hounds named Fennell and Quill (the researchers were so kind as to thank them in the acknowledgments of their paper, is how I know), went on the hunt. Sure enough, the dogs were able to track down more strongyle-infected birds than the dewormed ones. Like, *way* more. The hounds found only six birds with a low worm burden, versus thirty-seven with a high worm burden. The next year, it was nine versus twenty-nine.

You see, a moorbird with a gut full of strongyle starts smelling weird—and who could blame it, really. And that's very dangerous for a prey item out on the open plain. This is a smell our noses can't discern, but for a predator like a dog or fox, it's an invitation. But was this strategy selected for in the strongyle worm? Meaning, is making its host smell an adaptation to boost its chances of transmission? Well, no. The opposite, really—it lives a direct life cycle in the moorbird alone, as Janice Moore points out to me. "Now, in this case, this is not good for the parasite," she says. "Its only host is the grouse, so it dies when it gets eaten by a dog. But on the other hand, we know that it definitely intensifies the smell of the red grouse, which otherwise is relatively invisible to dogs." The lessons being: For one, not everything that seems to be a manipulation to benefit the parasite is in fact a manipulation to benefit the parasite. In the case of the strongyle worm and the moorbird, shit just happens. And two, the strongyle worm makes it clear that parasites—especially ones that live in intestines—can transform the way their hosts smell, whether it benefits the parasite or not.

And smell, Moore reckons, is where we should look (or smell, I guess?) if we want to find parasitic manipulations beyond our umwelt, because smell was probably the first sense on Earth. "Many scientists think that the ability to sense chemicals was perhaps the first sense to evolve—for the simple reason that even in very early, early, early organisms, the minute they started metabolizing something, they gave off chemicals," Moore says. "And then if another organism was hunting for a live dinner—that is, something that was metabolizing—natural selection favored the hunter that could pick up on those chemicals." It's an ancient system just asking for exploitation by parasites that can manipulate their hosts' smell for their own benefit—to end up in other critters' stomachs to complete their life cycles, for instance.

On the flip side, some zombifiers may be messing with how their hosts sense their world, with smell or otherwise. Remember that there's evidence to suggest that the acanthocephalan worms are using chemicals to make their amphipod hosts see light as darkness and darkness as light in order to steer them into position. Also remember that another acanthocephalan makes its host more attracted to the smell of predatory fish. Are other parasites heightening their hosts' sense of smell to make them better able to find new hosts to infect?* Really, there's no telling how many strange interactions of odors between parasites and hosts and predators have evolved over the millennia and have since left the Earth. Even those interactions that survive today elude science because humans are poorly equipped to detect them.

Save for the tale of the strongyle worm and the moorbird, a powerful manifestation of the invisible umwelten all around us. "When you think about all the ways that animals operate in the world," Moore says, "and all of the things we are insensitive to, parasites are doing a lot that we're not getting." We're biased, literally ignorant of what's right under our noses. Not only does a whole world of smell confound us, but so, too, does the tick's world of vibrations, as does the world of ultraviolet light that birds and bees can detect. Even then, the senses we *do* have may not be powerful enough to detect the subtlest of parasitic manipulations. We're lost in our own actuality of the cave, and it's very dark indeed.

## Breaking the Sound Barrier

Speaking of—there's no better illustration of the depths of human sensory hubris than the bat. Consider the tale of how scientists

---

*As is the case with your typical zombie of lore. While a zombie's eyesight might not be so great, usually its sense of smell is heightened. Presumably this helps it sniff out human hosts at night, when those hosts are most vulnerable, what with normal humans relying mainly on sight to sense the world. This also gives cinematographers an excuse to set scenes at night for added creepiness.

at first so arrogantly bungled their understanding of how a bat experiences its world. Well, one guy, an Italian priest by the name of Lazzaro Spallanzani, got it right in the eighteenth century. But his idea was so radical—bats navigate with a sense humans can't fathom—that one of history's most famous scientists laughed it into the ground without a shred of his own evidence to bury it.

Spallanzani was, again, working in the 1700s, so please forgive his experimental transgressions here. But he had a hunch that bats don't need their peepers to find their way through the darkness, so he got some bats and removed their ability to see,* then had them fly around a room. And sure enough, blinded bats could navigate as well as sighted ones. But, curiously for us modern folk who know full well that bats can "see" with sound, when Spallanzani sealed the bats' ears, it didn't have as much of an effect on their navigation as you or I might expect. Something with the experiment had gone awry. So Spallanzani, already incredulous that bats could even sense their world with sounds that the priest himself—a mighty human—could not hear, postulated that the creatures must be using some sort of sixth sense. That is, until a Swiss scientist by the name of Louis Jurine repeated Spallanzani's experiments and this time found that yes, deafened bats *did* have a hell of a time getting around. Jurine even managed to convince Spallanzani—a mighty human—as much.

But not the mighty human Georges Cuvier, one of the most

---

*I'm being purposefully vague here, lest I ruin the mood of a book that has hitherto avoided the topic of scientists blinding bats. Let's just say Spallanzani's methods weren't humane, and that if you absolutely must know you can consult my bibliography. The author you're looking for is Galambos.

Why do I feel like that still ruined the mood.

towering scientists of the day—or of all time, really.* Cuvier subscribed to the touch hypothesis, that a bat uses its hypersensitive wings to feel its way through the minute disruptions of air around it. Because why on earth would a bat be able to hear something we humans could not? (By this reasoning, though, the bat could therefore feel something we humans could not. So what's the difference, really.) And even though Spallanzani—again, not endorsing this—tested that hypothesis by coating bats' wings in varnish and showing the animals could still get around fine, if somewhat awkwardly on account of being varnished, the famous Cuvier's opinion took flight instead. No matter that Cuvier provided no bat experiments of his own.

"The touch hypothesis reigned for almost 150 years," Moore says, "until the technology that allowed us to actually hear bat vocalizations caught up with what bats had been doing all along." In the 1930s, Harvard scientists invented the parabolic ultrasonic detector, proving that Cuvier was wrong and that Spallanzani was on to something, however unsavory the priest's approaches may have been.

So not only humans, but *scientifically minded* humans, were so wrapped up in their umwelt that they couldn't comprehend that another animal might sense the world in an unhuman way, even when the evidence clearly suggested otherwise. "So that is

---

*Cuvier is most celebrated for breaking it to the world that God really has it out for some critters. The idea of extinction was tough for folks in the 1700s to comprehend, what with the higher power being a supposedly decent guy. But what Cuvier showed was that in the history of life on Earth, plenty of species have come and gone. He did this largely by comparing the bones of fossil elephants to modern elephants, showing that for one, they're distinct, and two, if these fossil elephants were still alive, surely someone would have spotted one. After all, losing track of a thirteen-thousand-pound species isn't quite as easy as losing your car keys.

the most convincing story of what happens when we don't pay attention to umwelt," Moore says, "and how we are so trapped in our own sensory world that we're willing to ignore the fact that deafened bats crash into things."

It's not just the bats that embrace a world of sound we can hardly fathom—sound pervades the world of the zombifiers as well. Take, for instance, the fact that the wingbeats of a certain male midge sound dramatically different under the influence of a nematode. Mind you, the scientist who studied this in the early 1980s, Wolfgang Wülker, didn't do it with the ears that Momma Wülker gave him. He used a microphone and something called a Pitch Computer PC 1400 (I googled it and got a lot of pictures of timing belts, for what it's worth) to determine that the tone of male midges plummeted from an average of 360 hertz to 230 hertz following infection. Weirdly, the latter is about the same pitch that female midges produce: 205 hertz or so. Also weirdly, unlike males, females infected with the nematode don't see much of a change in their sound. It seems the nematode is actually manipulating the shape and size of the male's wing, which in a normal midge is smaller than a female's.

A midge begins life as an egg that its mother deposits on the surface of a lake. The egg sinks to the bottom, where it hatches into a larva, which burrows into the mud and consumes organic debris. It's at this point that the nematode invades its body and feeds on its juices. As the midge transforms into a pupa, the worm messes with a male's development, feminizing it. The pupa then floats to the surface and emerges as an adult.

Now, males and females have different wing shapes because boys track down the girls by locking onto their distinct tone, swarming at the water's edge to mate. But worm-ridden midges of both sexes don't swarm with their healthy counterparts. Like a zombie clique, they form their own cloud over the lake. All gathered up in time and space, this is where the nematodes make their move. They bore out of their hosts' abdomens and drop into the water, mating and producing their own young. Perhaps by guiding the mating midges to a different part of the lake, Wülker hypothesized, the nematodes are better able to disperse themselves by finding newborn midge larvae to attack.

That manipulation would be fairly obvious. But not the wing tone. Why tamper with the male's wings, and therefore sound? Our man reckoned a bigger wing might help the male midge stay aloft with a load of worms weighing it down, very useful for a parasite with a vehicle to keep in working order. He couldn't figure out, though, what purpose the change in tone might serve. Really, it might not serve any at all—it could be an inconsequential consequence of a modified wing (though do keep in mind how suspiciously close the parasitized male's tone is to a normal female's). But that he couldn't parse it is exactly the point: We humans are equipped with ears, but that doesn't make us equipped to understand the umwelt of the midge. It took a Pitch Computer PC 1400 to even study the insect's different tones at all.

So scientists know a worm can change the tone of a midge's flight, and they know why the tone changes, and they know vaguely how the worm effects that change. But because they're struggling in their sensory cave, they don't know why tweaking the tone would benefit the parasite, or if there's any purpose to it at all, for that matter. And that's one conflict between two of the 9 million estimated species on this planet. Even if the worm isn't changing the tone of the midge for some purpose, you better believe another parasite out there is messing with its host's noises for nefarious means. Finding it, along with any number of other manipulators that mess with the smells we can't detect, means emerging from our sensory cave.

I understand it's a tall order. We've spent millennia doing our thing, seeing in a narrow range of light and hearing in a narrow range of sound, just as the tick instead concerns itself with vibrations and smells. But now—by getting over ourselves and admitting that we live in an objective world with far more stimuli flying around than we can detect with our senses—we have the opportunity to explore a strange new world of perception.

And who better to lead us there than the zombie ants.

# 8

# The Great Hacking of the Umwelt

***It's hard as hell to understand how another creature senses its world. Even harder? Hacking into its umwelt to ruin its life and complete yours.***

Murder is behavior unbecoming of butterflies. They're just so beautiful, so elegant, so delicate. Butterflies flitter about and sip nectar and land on puppies' noses, for Pete's sake. They don't murder—they *can't* murder.

That is, while they're butterflies. Because before the various species of the lazily named "large blue butterfly" become large blue butterflies, they become butchers. Adult females will track down colonies of *Myrmica* ants and lay their eggs nearby. The eggs hatch into caterpillars (butterflies are so magical we can't even call their larvae *larvae*—they're *caterpillars*), which infiltrate the colony. Then, depending on the variety of large blue, the caterpillars do one of two things, neither of which ends well for the ants: They'll brainwash the workers into feeding them or into letting them devour the colony's larvae. Not exactly normal conduct for a caterpillar, which in almost every species resigns itself to eating plants. And it's a mighty risk, trying to con feisty ants into raising you. But the rewards are huge: Not only does the caterpillar gain shelter, but it indoctrinates an army for protection as it develops into a butterfly that'll maybe, just maybe, land on a puppy's nose one day.

Given all the attention ants have drawn to themselves—from zombifiers like *Ophio* and the brainworms, not to mention all the other parasites that don't bother with manipulation—they'd do well to tread carefully. So what ants have developed is a sophisticated alarm system to keep intruders out of the colony, relying primarily on smell to differentiate friend and foe. Some species will even supplement their communication with sound, rubbing an organ on their abdomen to produce chirps. So if you've got a mind to infiltrate and exploit an ant colony, you might have two complementary lines of defense to penetrate. And as an extra hurdle for our caterpillars, the ant chirps aren't uniform in the colony—they differ between worker and queen. But that, oddly enough, is also a critical flaw the caterpillars have evolved to exploit.

Now, as I mentioned, these caterpillars employ two different strategies. The predator variety eludes detection by mimicking the scent of the ants and slipping by the workers, making its way into an isolated cell of the nest. From this hiding place the caterpillar from time to time squirms into the brooding chambers, devours some of the colony's own larvae, and squirms back. Any adult ants it might encounter along the way don't seem to take issue with this. More devious still is the cuckoo variety (named for the cuckoo bird, which lays its eggs in other birds' nests—more on them in a bit). It, too, mimics the colony's scent, but it's more brazen. Sitting among the colony's own larvae, this caterpillar demands the ants feed it by regurgitation, a process known more classily as trophallaxis.

All the while, both varieties are calling like their hosts, rubbing a minuscule tooth-and-comb organ like a fiddle. When a caterpillar calls out, the ant workers gather around, tapping it with their antennae or standing motionless atop the parasite—the exact posturing you'd find with workers attending a queen. And indeed, the caterpillar's calls more closely resemble those of a queen than a worker. So the intruder appears to gain the acceptance of the colony by mimicking its scent, then uses the calls to rise up the ranks to effectively become royalty, even though it looks like a lowly larva. And the workers fall for it hard. In the event of an emergency, they'll rescue the caterpillars before their own larvae, and even slaughter and feed their young to the parasites if food reserves are running low.

But this tale gets all the more improbable with the introduction of a third actor. The female large blue butterfly has to somehow find the right colony, for her young can mimic *Myrmica* ants and *Myrmica* ants only. And the oregano plant, of all things, is happy to help. When ants dig under it and tear up its roots, oregano releases massive quantities of the compound carvacrol, which is toxic to most ant species. But not *Myrmica*. That gives the ant an advantage, because it can take up residence under the plant without worrying about competitors moving in.

Problem is, that's a flare for the mother butterfly circling overhead. She preferentially chooses oregano expressing high levels of carvacrol—eight times as often, in fact—laying her eggs on the plant. The eggs hatch into caterpillars, which feed on the oregano's flowers, then drop to the ground. A worker picks up an ant-scented caterpillar, thinking it's family, and carries the fiend into the colony.

It's a poor decision: The caterpillar gets only 2 percent of its biomass feeding on the plant and the other 98 percent feeding on ant larvae. And more often than not, with the help of a few more caterpillars, it'll wipe out the colony by consuming the next generation. So in the end, oregano has weaponized a butterfly, and a butterfly has weaponized oregano. The caterpillar may feed a bit on its flowers at first, sure, but by annihilating the ant colony that's tearing up the roots, the caterpillar helps the plant thrive. Thus a resistance to carvacrol, which bestows *Myrmica* with an edge over other ants, can also be its downfall.

But this tale gets even more improbable with the introduction of a *fourth* actor, itself a manipulator. A certain wasp also infiltrates the colony, not to target the ants, but to use the caterpillar as a larder for its young. That means getting through the workers that are now under the distinct impression that the caterpillar is one of their own. So the wasp brings along some additional chemical warfare, in the form of a cocktail of compounds it releases, some that attract or repel the ants and some that induce aggression.

And so the ants close in and make contact—and go berserk. Tainted with the foreign pheromones, they tear off and attack their comrades, transferring the secretion to any ants they touch. As sisters rip sisters to pieces, the chemicals spread through the colony like a virus, immobilizing up to three-quarters of the ants.* Amid the chaos, the wasp strolls up to the caterpillar, injects it with an egg, and strolls out of the colony. The wasp larva consumes the caterpillar from the inside out, thus undoing the parasite that would be queen.

The killer caterpillar and its attendant wasp may not seem at first to belong to the league of zombifiers, but hear me out. They don't invade the body, no. They don't inject mind-hacking venom, they don't manipulate muscles, and they don't lodge in the brain. Because they don't need to: Their chemicals work from afar to remotely control their hosts. They're behavioral manipulators, using trickery to convince the host to abandon its own self-interest for the benefit of the parasite.

These are the hypnotists. For many of the ten thousand species of so-called social parasites, which exploit colonies of ants or bees or termites, hacking into the system is about hacking straight into the senses—into the umwelt—not physical manipulation. It's

---

*Which may remind you of any number of zombie movies in which a virus rapidly spreads through a population, instantly turning people into maniacs until hordes of the undead are sprinting through the streets. Yeah, not so much possible with a virus, but with the ants' fast-acting pheromones, definitely. The effect is temporary, but every bit as powerful as the manipulative zombie virus of lore.

about brainwashing not a single organism, but the *super*organism that is the colony. Which means the social parasites are occupying two complementary universes at once: the umwelt that their species would experience as a caterpillar or wasp to track down and exploit their victims, and the unique umwelt of the colony. That's no trivial feat, but if they can pull it off, they'll win an army of zombies to use as playthings. And maybe along the way they'll help pull us out of our sensory cave.

## Let's Not Necessarily Get Physical

The large blue butterfly caterpillars are far from alone in their exploitation of ants: A quarter of all butterfly species rely on the insects to complete their life cycles. This includes our predatory and cuckoo caterpillars that infiltrate the colony, and other varieties that more gently manipulate workers with bribes. And they'd do well to. Caterpillars are, after all, sluggish and helpless bags of flesh with hordes of enemies to worry about.

Take, for instance, the Japanese oakblue butterfly. Like many caterpillars, it secretes a sugary substance that ants go gaga for. So instead of butchering the larva for its meat, the ants lap up the liquid—and keep coming back, like farmers tending livestock. And like any good farmer, they'll attack other critters that dare molest their prize. Thus both parties benefit from the relationship, the ants gaining energy and the caterpillar gaining loyal little bodyguards. This isn't parasitism, then, but mutualism—with a little coercion thrown in the mix.

The oakblue takes this one step further by literally doping the ants. Well, if we're being literally literal, I suppose it's *reverse*-doping the ants: Open up the brain of an ant that's been feeding on the caterpillar's secretions and you'll find its dopamine levels have crashed—and for ants, dopamine is pivotal in regulating locomotion and aggression. So individuals habitually feeding on the secretion are more sluggish, yet more aggressive, meaning they're more likely to stick around the caterpillar and get scrappy. Think of it like a wasp's venom that turns its host into a bodyguard, only the ants are all too happy to fill their bellies with it.

And the caterpillar doesn't stop the manipulations there—it will inflate tentacle-like structures on its body when it feels threatened, signaling its soldiers to attack the enemy. (Place ants that haven't yet fed on the secretion near a caterpillar and they won't leap into action when the tentacles pop out.) So the caterpillar is dosing the ants to broadly manipulate their behavior, then refining that behavior by ordering its zombies around with visual cues.

As if caterpillars weren't enough to worry about, ants aren't even safe from other ants. In the Swiss Alps, the *Teleutomyrmex schneideri* ant lives a confounding life as a parasite of the pavement ant. Unlike your typical ant, this extremely rare species has no worker caste, so instead the *Teleutomyrmex* queen clings to its host queen. Or several do, I should say—as many as eight will piggyback on a single pavement ant mother. That, unsurprisingly, immobilizes her, but the pavement ant workers don't find this to be worrisome. In fact, they'll groom the parasites, which appear to be releasing an attractant pheromone from glands all over their bodies. The workers will even feed the imposter queens, which is more than welcome: *Teleutomyrmex* is so degenerated in her physiology that she's incapable of taking care of herself—her mandibles are worthless and her brain is tiny. But that's all right, because the parasite is perfectly adapted, in her own way, to exploit the pavement ant queen. She's got a concave belly so she can better snuggle her host, for instance, and huge claws to get a good grip. And, most important of all, she produces pheromones that brainwash the workers into pampering her.

Other ants that parasitize their own kind aren't so gentle as *Teleutomyrmex*. A *Polyergus* queen, for example, is incapable of founding a nest of her own, so she enslaves colonies of *Formica* ants. As she penetrates the nest, the queen releases a "propaganda" pheromone that calms the workers, then makes her way through the colony unmolested. When she gets to the royal chambers she attacks the true queen, repeatedly biting the head and abdomen and licking the wounds to assume her scent. And this is almost certainly her aim here: Present the usurper with a dead and therefore motionless (as is typical of dead things) queen and she'll bite the bejesus out of her as well, suggesting *Polyergus*'s pursuits are as much about assassination as they are about adopting a chemical disguise. The true queen eventually expires, but her subjects don't avenge her. (Another species of parasitic ant, known as *Bothriomyrmex decapitans*, employs a more . . . targeted approach, clinging to the host queen's back and slowly sawing off her head.) In fact, under the spell of olfactory propaganda, they'll begin grooming their new queen and caring for her eggs. The ruse is so effective that if you pick up this queen and drop her in a second *Formica* colony, the workers will immediately accept her as their own. She's a different species, but to the plebs, she'll always be queen mother.

## The Inherent Poetry of Bumblebees Slaughtering Each Other

Bumblebees, too, suffer manipulation from afar. The *Psithyrus* cuckoo bumblebee queen is also incapable of founding a colony of her own, so she conquers the nests of other bumbles. This unfolds rather more violently than is the case with the ants and caterpillars, as a team of scientists discovered in the late 1970s. (I can't resist quoting their paper in the coming paragraphs. It's not laziness—this is a lovely brand of scientific writing you don't see too often anymore.)

The researchers collected bumblebees from the garden of an old fortress in the Netherlands (see, poetic already), and used them to establish a colony of forty workers in a box. Then they introduced a *Psithyrus* queen to see what would happen. At first the workers merely inspected her. "Another worker in the outer compartment lifted her middle leg on meeting the *Ps.*, rolled over onto her back, and defecated." The queen continued to penetrate the nest until—three minutes after her first run-in with a worker—another worker grabbed her and held tight. The cuckoo, though, "got rid of the worker easily and started walking through the box again."

Soon came the first signs of alarm. The cuckoo queen approached a group of five workers, which got a-buzzing. "After a few seconds the whole nest seemed to be alarmed and all the workers started to walk very quickly" around the colony. Then, the opening strike. "Some ten workers attacked the *Ps.* and balled her while trying to sting her." Others, though, were more interested in stinging one another. *Very* interested: In the end, 30 percent of worker casualties came not from the intruder, but from infighting. The finely tuned computer that is a bumblebee colony had begun to glitch.

Meanwhile, the colony's true queen remained calm. Until, that is, her own workers began attacking her—not that they let up any on the cuckoo. "The fights between the *Ps.* and the workers continued for 30 minutes and, at the end, 31 workers were dead.

The *Ps.* rested on the nest for several minutes and tried in vain to remove the head of a worker from her left anterior wing."

After her repose, the cuckoo again wandered the nest, grabbing any remaining worker she found and extending her stinger, yet stopped short of stabbing them, as if to intimidate. But the queen she left alone, the two sitting together on the colony's comb of larvae "without apparent signs of animosity." Four hours after the cuckoo first infiltrated the nest, calm prevailed and continued for several days—until the cuckoo made her move, pushing the true queen off the comb, chasing her out of the nest, and destroying her eggs and larvae.

Comfortably in power, the cuckoo began laying her own eggs. But so, too, did the colony's workers (unlike with the queen, their unfertilized eggs can only result in male offspring). This did not suit the new queen. "The *Ps.* chased the laying workers off their brood with great ferocity, and 11 workers were killed in these encounters." Then she ate their eggs and killed their larvae. This, in turn, did not suit the workers. They began harassing her, buzzing about, until the cuckoo got fed up and walked away. Two days later, "the workers became ferocious," forcing the imposter off the comb and destroying some of her eggs and larvae. Still she stuck around the nest as the harassment intensified. "She was always surrounded by several workers, who buzzed near her head and eventually started to pull her legs and wings." Finally, a month after she'd infiltrated the colony, she left the comb and died a day later. Two weeks after that, nine of her children emerged—three females and six males—and fled the nest their mother had conquered.

The female cuckoo bumblebee placates her hosts by mimicking their pheromones, at least initially, then probably adopts their scent as she lingers in the nest. Only when she's fully absorbed it will she chase off or kill the resident queen. The ruse isn't always perfect, hence the intermittent battles, but it's good enough.

But what about male cuckoo bumblebees? They're born in a foreign nest, just like their sisters, so how do they manage to escape without their caretakers stinging them to death? The answer is also pheromones, and in fact the males' trickster chemicals may help the females get into host colonies in the first place. Inside a male cuckoo bumblebee's head is a huge gland that pumps out sex pheromones. That's meant for the females of his own species, of course, but the mixture also contains compounds that repel host bumblebees. When cuckoo bumblebees come together to mate, the females pick up this scent and carry it with them into the host nest. The females are producing their own compounds to mimic the smell of the hosts, sure, but the extra repellent they get from the males is like a good spritz of perfume. So by producing this stuff, the males can escape their foster homes unscathed *and* give their mates a better chance of infiltrating another colony. Think of it like a magical bouquet of flowers, which would be romantic if it weren't for all the death involved.

We humans don't traffic in this sort of thing. Smell just isn't that important for our umwelt, where sights and sounds reign. For the ants and bumblebees, though, pheromones are both an indispensable form of communication and a liability, an entry point for the parasites crafty enough to break the code. Evolve a counterfeit pheromone and you can zombify the superorganism that is the colony, sending workers into a tizzy while you ascend the throne.

While I'm fully aware that I can't just stick my face in an ant nest and discern the colony's unique smell, knowing that the zombifiers are leading scientists through such a foreign umwelt is reassuring. Not that I want to see ants and bumblebees suffer, but the exploitation of their colonies provides a fascinating journey out of our own umwelt and into another, a whole world of pheromones that scientists are decoding little by little. Which is not to say that our more familiar umwelt of sights and sounds is immune from exploitation. No, that code is also worryingly easy to crack, no chemicals required.

## That's a Nice Family You've Got There. It'd Be a Shame If Someone Parasitized It.

Imagine, please, that you're the parent of a baby girl. And let's say you're what we might call a "good" parent, in the sense that you feed the child so she doesn't starve and clothe her so she doesn't freeze. Now imagine that another new parent in the neighborhood, having heard that you're a "good" parent, breaks into your house while you're sleeping and deposits their own baby girl in your crib.

Upon discovering such an addition to your family, would you:

A) accept the rascal as your own
B) fail to notice you have an extra kid
C) take the intruder to the nearest fire station because you have enough responsibilities as it is

If you chose option C, it's probably because you're a busy person, and there's nothing wrong with that. If you chose A or B, you're probably one of the many poor species of bird that fall victim to the deadbeat-parenting antics of the original brand of cuckoos. These devious avians lay their eggs in the nests of other birds like warblers and wrens and waxbills, which adopt the parasitic young as their own. But the cuckoos aren't manipulating their hosts with smell like the zombifiers that exploit social insects—these are illusionists. It's a convoluted process that involves gaining access to a nest, depositing an egg the owners don't grow suspicious of, and hatching a chick that doesn't strike the hosts as, you know, a different species.

There are two types of cuckoos in this world. (Well, three, really. The majority of species actually raise their own young, but that's none of our concern.) We have the evictors, whose chicks hatch earlier than the host's chicks and who then—like Atlas shouldering the spherical sky—roll the rest of the eggs onto their backs and out of the nest. That or the evictor waits until everyone is hatched, then uses its sickle-sharp bill to stab its adopted siblings to death, at which point the parents do the evicting by tossing the bodies out of the nest. Either way, in the end the cuckoo chick is the sole survivor, and the host parents feed it as if nothing happened. That means the imposter doesn't have to compete with any adopted siblings, so it consumes every scrap of food its newfound guardians bring it.

The second type of cuckoo is less bloody in its parasitism, but necessarily craftier. These nonevicting chicks grow up alongside the host chicks, all the while begging for food. If the true chicks are lucky, the parasite will steal some grub, grow to maturity, and fly away. But the cuckoo can get a bit . . . carried away with things. It might be so insatiable that its foster parents neglect their own kids in order to feed it. The other chicks either starve or the cuckoo—ballooning to ten times the size of the *adult* hosts—crushes them to death.* But why do the parents see nothing wrong with continuing to care for the giant that's massacred their family? Well, because it's too risky not to.

Really, it's not like host birds aren't aware that cuckoos are bad news—they do everything they can to keep the cuckoos away in the first place. They'll form a neighborhood watch, several individuals posting guard and calling out when a cuckoo starts snooping around. Should the parasite close in, the birds will mob it and hope that violence is indeed the answer. Certain host species even seem to have evolved to build domed nests with tiny entrances that keep out the cuckoos, which are fairly large birds. These are all valiant efforts that the cuckoo has counterevolved ways to circumvent. Cuckoos will work as a team, for instance, the male pestering the victims while the female slips in and lays her eggs. Some even look nearly identical to the birds of prey they share a habitat with, right down to their patterning and shape, thus terrifying their marks into submission.

Should a female cuckoo infiltrate a nest and lay her egg, she needs to make sure the host can't tell the difference between it and the others already in there. So a cuckoo that specializes in parasitizing a particular species of bird will lay nearly identical eggs. If the host's is a solid blue, the cuckoo lays a solid blue. If it's mottled, the cuckoo's is mottled, too. Even if the host parents catch the cuckoo invading the nest and laying an egg, the costs are too high to do anything but submit: It's either throw all the eggs overboard and start anew or pray to the avian gods that the cuckoo they're sitting on isn't the evicting variety.

---

*As a child of the 1980s and a fan of the "Cuckoo for Cocoa Puffs" TV ad campaign, in which a charming cartoon cuckoo bird does indeed go cuckoo for Cocoa Puffs, I am having confusing feelings given all this.

So typically species that fall victim to cuckoos deploy a countermeasure, laying eggs that don't vary much within the clutch. This makes it easier to spot an imposter than if the mother laid eggs of, say, different intensities of speckling. And by combining color and patterning, hosts can create a verification system of sorts, "much as the complicated markings on banknote watermarks render them more difficult to forge by counterfeiters," as one pair of researchers once put it. (Some species of cuckoo leapfrog all that by hiding their eggs in a host's domed, shadowy nest. The legitimate eggs in there are white, but the cuckoo lays olive-green ones, which the host parents don't notice against the darkened background of twigs.)

Which brings us back to the conundrum of the umwelt: Birds don't see the world like we do. Yes, scientists can show that host birds identify and evict cuckoo eggs that look different from their own. Or, I should say, the eggs look like how we *assume* the birds think they look. The problem is, birds can make out patterns and colors in the light that's visible to us, but they also see the ultraviolet light we can't. Which might explain why sometimes host birds don't single out cuckoo eggs that look laughably different. A human can easily tell apart an egg of the red-chested cuckoo, for instance, and that of a bird it parasitizes, the Cape robin, but a Cape robin can't. That's because the two eggs reflect UV light in strikingly similar ways. The Cape robin is no dolt—it's just stuck in its own umwelt. It's we humans who are shortsighted for assuming it must see as we see.

We are (okay, maybe I am) also assuming that smell isn't a factor here—maybe the cuckoos *are* working with manipulative scents like the social parasites that torment ants. Naturalists have long thought that birds can't smell worth a damn. But that assumption is crumbling, thanks in large part to tampons. In

1991, ecologist Gabrielle Nevitt set out to prove birds can smell by dipping the feminine hygiene products in fragrant fish compounds and flying the things on kites off an Antarctic research vessel. The stunt attracted so many seabirds that Nevitt had to reel the kites back in, fearing the creatures would get tangled up in the lines. Which is all to say, perhaps cuckoos and their hosts are indeed interacting via scents on top of visual cues and sounds. It's just that we're trapped in the human umwelt, and there's not much we can do to escape it and fully appreciate the prevalence and sophistication of mind control among birds, or other zombifiers for that matter.

Matters grow all the trickier for the host parents once the cuckoo chick hatches, for they're no longer scrutinizing an egg that's static in appearance, but a youngster that's changing day by day. And many species of cuckoo beautifully mimic their hosts right out of the shell. The little bronze-cuckoo sports awkward patches of down on its otherwise shorn body, like its host, the equally hyphenated large-billed gerygone. Other species have coevolved to closely match the color and size of the chicks they share a nest with—but not for long. Cuckoos will grow at a rapid clip, quickly dwarfing the host chicks as their foster parents work overtime to feed them. Still, the dupes never stop providing.

That's due in large part to a unique facet of parent-chick communication: gape patterns. Nestlings make it known that they're hungry by opening wide and chirping, a display made all the more emphatic in some species by complex patterning inside the mouth—black and white dots and C-shaped flourishes. Cuckoo chicks, too, will deploy these markings to complete their ruse. This is especially important for the evictor varieties, which by their assertive nature end up alone in the nest without legitimate chicks to help them tell the parents they're hungry. One species,

Horsfield's hawk-cuckoo, solves this by flashing a bald patch under its wing, tricking the parents into thinking there's another mouth to feed. It's so effective that they'll try to cram food into the chick's bald spot. Evictors like the common cuckoo will also employ sound to trick their foster parents into thinking not only that they're hungry, but that there are still other chicks in the nest, speeding up their begging call to mimic the cacophony of several nestlings. And it works: The parents will feed the solitary cuckoo as much as they would a whole nest of youngsters.

## Bastardizing the Umwelt

So here we have a parasite that's evolved into a master manipulator of the avian umwelt. Like any good zombifier, the cuckoo hacks its host's brain to abandon the organism's own self-interest in favor of the parasite's. It isn't exploiting smells (well, at least that we know of) like the social parasites of ants, but sights and sounds. A female first dons the disguise of a predatory bird, terrifying her marks into submission. She lays an egg that mimics the host's, both in what we'd call "visible" light and ultraviolet. The egg hatches into a young cuckoo that looks much like its adopted siblings—if the cuckoo hasn't tossed them out of the nest already. Right on down to the mouth and chirps, the cuckoo is a stunning fraud.

The cuckoo hypnotizes parents into doing something very unfortunate: Giving up the chance to pass down their genes to the next generation in favor of doting on the enemy. It's manipulation from afar, yet it's just as effective as the chemical zombification of the wasps or the nematomorphs. Getting another creature to do your bidding isn't always about physically tweaking the brain, but about hijacking the umwelt, about fooling your host into becoming your servant—hack the senses, and you hack instincts. It shows how powerful yet precarious the umwelt can be: The host bird is as handicapped by its umwelt, unable to sense beyond what its biology affords it, as the cuckoo is emboldened by its own umwelt. The cuckoo "knows" the sensory limitations of its victim, and it exploits those limitations brutally.

Sometimes, though, the dupe doesn't get the hint, in which case the manipulator makes it an offer it can't refuse. The magpie, for instance, doesn't take too kindly to the great spotted cuckoo laying eggs in its nest, so it'll destroy the imposters. This the cuckoo certainly doesn't appreciate. So the parasite checks in on the host nest and destroys the entire clutch if the magpies have been putting up a resistance. In other words, the cuckoo gives the magpie parents a simple choice: Accept my egg and spend the time and energy raising my chick, or I'll take something very dear to you. And indeed you'll find that some magpies will get the hint after the cuckoo smashes up the nest, accepting the parasite's egg the second time around. It's another manifestation of the mafia hypothesis we saw with the nematomorphs—a host deciding that cooperating with the parasite is in its best interest. For the cricket it's about leaping into water to offload the worms that would otherwise kill it, and for the bird it's about weighing the cost of laying more eggs, a tremendous burden of energy and bodily resources.

And all because the bird is stuck in its umwelt, because while this umwelt gifts the creature with the ability to navigate a world of countless potential stimuli in its own unique way, that makes it vulnerable. If a parasite manages to hijack that umwelt, the victim is blinded to the objective reality that every species on Earth inhabits, all because long ago its kind evolved to rely on seeing a particular range of light and hearing a particular range of sound.

Just think about how easy it is to trick your own umwelt, one not dissimilar from that of the birds. Optical illusions hack our visual system. More subtly, colors can play with us on both a physiological and a cultural level—evidence points to blue light having a calming effect on humans, while marketers exploit blue to brand something as masculine, as opposed to pink representing femininity.* Auditory illusions, too, mess with our ears. Use a computer to garble a human saying "I believe in the concept of free will" and a listener will at first have no clue what's being said. But play the ungarbled version, then the garbled version again, and suddenly you can hear the nonsensical words because your brain is applying new information to unscramble the puzzle. Even smells can affect the mind in powerful ways—think of the classic Realtor-using-freshly-baked-cookies-in-an-open-house trick. Conversely, our brains are wired to recoil at the stench of rotting food, for obvious evolutionary reasons. Thus simple manipulations of our umwelt work directly on the brain.

The more zombifiers I get to know, the less I trust my mind. I'm not worried about a wasp stinging me in the brain or a fungus growing in my body, but about the *implications* of what the parasites do: hacking pure biology with known hormones and neurotransmitters. And it's the manipulators of the umwelt that worry me the most. We're biological beings with sensory weaknesses, like the birds that fall victim to the illusions of the cuckoo or the ants that are undone by olfactory propaganda. Our brains

---

*I spent a rather unhappy nine months as a copy editor at a dental marketing firm that produced mailers asking people to visit their local tooth professional. They were colorful brochures, yet were devoid of red. The thinking went that red would remind people of blood and conjure memories of visits to overly aggressive dentists.

Nine goddamn months of my life.

have evolved over hundreds of millions of years to simplify a world of dizzying chaos into something more manageable—our own personal umwelt. Something happens, we detect it, and we react without half a thought. If you didn't have such an autopilot system, you'd have a meltdown pretty much immediately. So how much are you in control, really?

If you haven't been worrying about your own brain by this point in the book, now's the time to start.

9

# The Brain-Hacked Mouse That Wore a Funny Hat and Destroyed the Notion of Free Will

***This is going to be tough to come to terms with, but bear with me: There's no such thing as free will, for you or any other creature.***

For something so fluffy and adorable, this mouse sure is an asshole. Adorable, I guess, save for the plastic crown glued to his head, from which a wire erupts and runs to a nearby computer, from which erupts the sound of his neurons firing. *Click... click... click... click... click.* Still, though, definitely an asshole, though it's not really his fault.

"So you know how you can train a rat to press a bar and get a pellet?" neuroscientist Annegret Falkner asks me in her NYU lab. She wears her wavy hair done up and a denim shirt over a red floral dress. "I trained mice to do that, but instead of a pellet they get to beat up on another guy for five seconds."

Stick his snout in the hole to the left, Falkner has coached him, and nothing happens. But stick his snout in the hole to the right and he breaks an infrared beam, summoning out of the sky a latex-gloved hand that lowers another mouse into the container. And not just any mouse—a submissive male our alpha is free to kick the daylights out of. This Falkner refers to as a social reward. "You can tell that he's learning because he pokes and he looks up, and waits, and looks at you," she says. "He's completely aware of what's happening." Mind you, it isn't just about our alpha getting aggressive toward any mouse Falkner lowers into the cage at random. By tripping the infrared beam, he's *requesting* a fight. He's a frat boy, filled to his eyeballs with vodka Red Bulls, leaving a bar at two a.m. And someone's glued a hat to his head.

When the alpha does pick that fight, he strikes with mystifying speed and ferocity, more like a rattlesnake than a mammal, biting the submissive male on the rump and giving him a shake. Bedding flies and the opponents' tails smack the plastic walls of the arena, as the submissive mouse tries to curl around and defend himself. But no luck. The alpha is bigger and angrier. The stooge struggles still, managing a feeble squeal, staring at the ceiling with tiny red eyes.

Then, deliverance: The latex-gloved hand grasps him by the tail and pulls him into the sky. "Basically," Falkner says, "the fight is always fixed."

In this lab, which features a hint of the musty scent of rodents and a view of New York City's East River to enjoy between mouse fights, Falkner gets critters mad. And she hasn't simply trained mice to battle—she's hacked their brains to turn the creatures into serious scrappers. The alpha males are literally hardwired to fight: Falkner's implanted fiber-optic cables in their brains, specifically in a region called the hypothalamus, which among many neurological responsibilities controls aggression. Shine a laser through the cable, and Falkner can turn a mouse hyperaggressive, to the point where it happily attacks not only other mice, but also inanimate objects like inflated gloves. "You can stimulate animals and actually get them into a state where they want to seek out aggression much faster, and much more frequently," she says.

This is optogenetics, the tampering of animal behavior not with drugs or electric shock, but with light itself—and a dash of virus. A jolt of electricity to the brain is, after all, far too slapdash. Far too . . . *Frankenstein*-esque. And mind-altering drugs take too long to kick in and wear off. Far too . . . inexact. Optogenetics is humans precisely manipulating other animals, what took the body snatchers millions of years to slowly evolve step by step. And by coincidence, Falkner's doing it a whole lot like the jewel wasp. Which means she's first gotta sting.

Falkner begins by getting a mouse loopy on anesthetics and loading him into a device of many dials, wrapping his mouth around a bit that delivers more anesthetics during the procedure. Next she gently inserts metal screws into each ear to keep him in place. With turns of the many dials, she precisely positions the rodent's head. "I give him a little haircut," she says, sheering away his dome, leaving the mouse with the fetching tonsure of a medieval monk. All the while the patient's unconscious body gently heaves.

The procedure itself starts with an incision lengthwise along the skull. Then Falkner lowers a drill down to the cranium, staring through a microscope to position a bit of extraordinary thinness with a turning of the many dials. (Do remember that we're talking not only brain surgery here, but brain surgery on an animal that'd fit comfortably in your hand, and whose brain would fit comfortably on your fingernail. And unlike the jewel wasp, Falkner can't feel around in the brain with her stinger.) Slowly, Falkner brings down the bit and penetrates the skull, producing a dusting of bone. She next replaces the drill attachment with one holding a needle of extraordinary thinness. This she inserts ever so carefully into the hole in the skull and continues gingerly through the brain until she reaches the hypothalamus at the base, where she injects her venom.

Like another of our wasps, *Cotesia*, Falkner has weaponized a virus, one loaded with a green alga protein called channelrhodopsin. When Falkner injects this virus into the mouse's hypothalamus, it breaks into individual cells—in this case neurons—as viruses are wont to do. (As a more everyday example, the flu virus is up to the same shenanigans, breaking into your cells and turning them into factories that produce more of the virus.) Here the weapon delivers its payload, which bestows the neurons with a special ability: light sensitivity. Falkner then removes the needle and threads her fiber-optic cable to rest above these corrupted neurons. The cable runs to a plastic hat, which she glues with dental cement to the skull. Attach the cap to a laser, and Falkner can fire light through, activating the hijacked neurons in the hypothalamus to turn the mouse into a bit of an asshole.

We might then ask: Well, why? Why turn a mouse into a frat boy? The answer is that none of this is about the mice, at least not in the big picture. Mice retain the evolutionarily ancient hypothalamus and, by virtue of also being mammals, so too do people. Better understand the roots of aggression in mice, and you can better understand the roots of aggression in humans. When you understand that, you might be able to one day control that aggression—though of course not with the same methods.

What Falkner's experiments show is that the brain is a predictable machine. The mysteries of human consciousness may persist, sure, but everything that the brain produces is rooted in biology, in tissue, in electrical signals. And that means—as every one of our body snatchers has discovered over evolutionary time—that the brain is easily hacked.

Oh, and it also means free will is a lie.

## Welcome to the Machine

Back in 1867, Mark Twain had an encounter with a Parisian swan like no other, a bird so pristine that it practically glowed. Unlike pretty much every other swan ever, it didn't attack him for no reason. It simply sat in shimmering water, craning its neck up and down, stretching to preen the feathers on its rump. Very peaceful-like, honking not a single honk. Which was all very impressive, considering the swan was by this time a century old.

What Twain met was not an animal, but a hand-cranked automaton of extreme complexity. Its feathers were made not of keratin, but silver. It sat not in a lake, but atop rotating glass rods that shone like sun-blasted water. All told, the magical creature—built by the appropriately named John Joseph Merlin in 1773—consisted of two thousand moving parts, with over one hundred rings making up that busy neck alone, all working in concert to create a robot that wowed even a cosmopolitan like Twain. "I watched a silver swan," he wrote in his travelogue *The Innocents Abroad*, "which had a living grace about his movements and a living intelligence in his eyes—watched him swimming about as comfortably and as unconcernedly as if he had been born in a morass instead of a jeweler's shop—watched him seize a silver fish from under the water and hold up his head and go through all the customary and elaborate motions of swallowing it—but the moment it disappeared down his throat some tattooed South Sea Islanders approached and I yielded to their attractions."

The Silver Swan was meant to impress, not serve as a metaphor, but let's take it as a metaphor anyway: Organisms are machines. They have predictable inputs and predictable outputs. A human operator inputs energy by cranking the avian automaton, and out comes a choreographed dance. Same goes for a real swan, as food goes in and what we know as generally swanlike behavior comes out—preening and chasing children and such. The animal is programmed for the ultimate goal of reproduction, and along the way the brain commands it to eat and sleep and defend itself from predators, all behaviors the swan doesn't "decide" to do. The machine exists to propagate the machine. It's a servant to its genes, which from the moment of inception take command of the animal, ordering it around until death, at which point the machine powers down. The creature's life, then, is largely predetermined, just as the Silver Swan obeys the gears and springs and hinges that cascade in the right way to produce a dance that a godlike inventor programmed centuries ago.

It's tempting to think of the brain as a computer that drives the cognition of a swan or a human or the victim of a parasitic manipulator, running on the program that is DNA. But let's not. A computer operates in relative isolation: It takes your input of clicks or keystrokes or taps and uses code to translate it into some kind of output, like images or sound. It has to wait for something to come along and stimulate it. But a swan doesn't work like this. It's always in communication with its umwelt, responding not only to active input like children invading its personal bubble, but more passive input like temperature fluctuations and light and darkness. A computer sits there passively, while an animal mills about, constantly sampling the sights and smells and sounds of the world around it.

So the philosopher Tim van Gelder has proposed a better (though more old-school) metaphor for cognition: the Watt governor. Think of a Watt governor like a figure skater doing that thing where they spin in place really fast. Give the skater a bowling ball to hold in each hand, and as the skater spins faster and faster, centrifugal force will make the arms extend—slow down and the arms fall back down to the skater's sides. In the Watt governor, those bowling balls are called flyballs, and they're attached by arms to a central spindle, which itself is hooked up to the shaft of a steam engine. The flyballs are also attached by way of a lever to a throttle that controls how much steam is flowing into the engine. So if the engine, and by association the spindle, gets going too fast, the spinning flyballs will extend with centrifugal force, pulling the lever that closes that valve, thus powering down the engine. In this way the parts of the system work in concert to automatically maintain a speed.*

This is how we have to think of the brain and cognition. We can equate the extreme complexity of a computer with the extreme complexity of the brain, sure, but that confines us. Like with the Watt governor, an animal is in constant communication with its environment. It does not make a calculation and power down when that calculation is complete. The brain, nervous system, body, and umwelt are one. That makes me sound a bit woo-woo, I know, but it's how things work.

So manipulative parasites aren't just hacking brains, but reengineering governors to fundamentally transform the ways in which their hosts operate in the surrounding world. The acanthocephalan turns light into darkness, convincing an amphipod to swim into the belly of a bird. The *Cotesia* wasp converts a normally peaceful caterpillar into a bodyguard that bashes the hell out of predators. And *Ophio* demonstrates the point best of all: The fungus may flood the ant's brain with chemicals, sure, but it's also growing throughout the muscles to pull the individual strings. Saying that this is merely a parasite hacking a computer

---

*As it would happen, Alfred Russel Wallace, in the panic-inducing paper he sent to Charles Darwin introducing his theory, used a centrifugal governor as a metaphor for what he saw as the inherent balance of evolution: Natural selection (though he didn't call it that) weeds out any weakness. "The action of this principle is exactly like that of the centrifugal governor of the steam engine," he wrote, "which checks and corrects any irregularities almost before they become evident; and in like manner no unbalanced deficiency in the animal kingdom can ever reach any conspicuous magnitude, because it would make itself felt at the very first step, by rendering existence difficult and extinction almost sure to follow."

ignores the fact that the host isn't merely a brain. That, and the fact that the host never had any free will anyway.

## The Whole-Body Model of the Great Big Sham That Is Free Will

I'd like you to close your eyes—wait, not quite yet—and think of something other than this book for however long you please. Anything. Put the book down if you must.

Okay, close your eyes now.

. . .

And welcome back. Now, why did you think of [insert whatever you thought of here]? Of all the things in the universe, all your knowledge and memories, everything surrounding you, that's the one thing that came up.

I hate to break it to you, but you didn't make the decision to think about that, because to do so would mean you thought about thinking a thought, which means you would have also had to think about the thought about thinking a thought, and then think of a thought of a thought of a thought of a thought, and you see where I'm going with this. That would be an inefficient way for a brain to operate. More importantly, I never told you to open your eyes again and focus them back on this page. You didn't make that decision—your brain made it for you.*

The notion of free will has troubled humans ever since the first philosopher first philosophized. The debate, simply put: Are we

---

*Neuroscientist Sam Harris's version of this thought experiment has you think of a city off the top of your head. You can try doing that as well, but unfortunately you're still not freely thinking of that specific place.

in control of our actions, or is that an illusion constructed by the brain, which—organized as billions upon billions of neurons—starts up at birth as an engine and plays out our lives one decision after another? It's a fine conundrum, considering it rests at the core of human society. Why prosecute someone for theft or assault or murder if that person didn't "choose" to do so, but simply acted out what their brain destined them to do from birth? Why bother with morality at all? Great questions that are growing more profound, considering neuroscience now suggests that your free will is indeed a sham.

The first solid neurological experiment to question the notion of free will came in 1985 from a scientist named Benjamin Libet. He wired up test subjects to an EEG machine, which monitors the electrical storms that rage when the brain processes thoughts, and asked them to press a button and make note of the time when they decided to do so. Imagine the subjects' unrelenting boredom sitting there jabbing a button, and then imagine their astonishment when the study revealed that their brains lit up with activity a half second before they were aware of their decision. Meaning the brain was making the choice when to press the button, not a "conscious" agent.

Subsequent studies confirmed the phenomenon, but over two decades later, a follow-up experiment to Libet's found an even more startling disconnect. This time subjects had the choice between hitting a button with their right or left hand. By monitoring their brain activity, scientists could not only see the brain making the decision seven seconds before the subjects realized it, but predict which hand they would use. Still another experiment a few years later tasked subjects with making a purely intellectual calculation instead of a mechanical one (like pressing a button), which we can assume was marginally more intriguing for the participants. In this study, an fMRI machine—which detects blood flow in the brain instead of electrical activity—caught the subjects making the decision to either add or subtract a series of numbers four seconds before they recalled doing so.

Now, we need to note that this kind of science is still in its early days—the technology for detecting brain activity is great and all, but how the mind works is still largely a mystery—and that deciding to press a button is wholly different from deciding to pull the trigger and kill someone. But study after study has suggested that the agency we experience when making choices is a lie. That may make you feel weird, and it may be scary for a society that champions personal responsibility. But it's in keeping with our conception of the animal as a governor, constantly reacting to the environment, making thousands of subconscious choices every day. Its genes built its body and its brain, and it's stuck with that biology from the day its neurons wire up until the day it dies.

I'm not saying our environment has no effect on us. No, exactly *because* we're governors, constantly interacting with our environments, do we change psychologically over time. Traumatic upbringings severely alter children, for instance, and PTSD works on a deep psychological level to change even developed brains. But what's transforming is a brain that is itself a product of DNA. And indeed animals are hardwired to cognitively adapt to their surroundings, which are in constant flux, because going through life as a static organism is asking for trouble. A cheetah, after all, may be born with a longing for meat, but its mother still has to teach it to hunt. For humans, learning is such a mighty force that even an adult brain can physically change as it loads up with information, a capability known as neuroplasticity.* But all the while, it's the underlying genes that are running things. Every single organism, no matter how well it can learn and evolve emotionally, is still a slave to its DNA.

Every single organism, that is, except the zombie with a parasite in its brain.

Consider the ant. It's born into a colony that it dutifully feeds and defends. That's because its genes tell it to—until the *Ophio* fungus comes along and overrides the marching orders of the host, turning the ant against the colony. In a sense, then, the fungus hasn't enslaved the host, but *set it free* from the confines of its genes. An ant's DNA wouldn't tell it to climb a certain ways up a tree and bite onto a leaf and hang upside down for the fun of it. The insect has better things to do. So the fungus releases the ant from its regularly scheduled programming, which of course is a short-lived release and a rather nasty one. But you and I, we'll never transcend our programming—no one will ever Manchurian Candidate us. We'll eat, drink, sleep, and reproduce like our brains tell us to. The neurons fire, and we follow the orders.

This makes me feel gross and liberated all at once. I hate the idea that I am in a way a zombie on autopilot—save for the yearning for human flesh. I hate the idea that I'm not in control. I'm a human, damn it, a being capable of complex decision-making. I,

---

*The most fascinating example being the Strange Case of the London Cabbie's Transforming Brain. Cab drivers here must pass a trial known as the Knowledge, memorizing some twenty-five thousand streets and twenty thousand landmarks. It can take up to four years. In that time the cabbies' brains physically expand, specifically a region called the hippocampus, which is responsible for memory and spatial awareness. And when they retire, their hippocampus shrinks back down. Use it or lose it, as the cliché goes.

for instance, must have made the decision to have a salad for lunch today because I respect my body. Which, alas, is untrue on two levels. First, I didn't consciously make that decision. Also, I don't really respect my body, which itself isn't really a decision to make, which is a relief. Yeah, it sounds like a cop-out, but boy does it help take the burden off of figuring out what to have for lunch.

This is, of course, a dangerous idea, verging on a kind of biological nihilism. If free will doesn't exist, then why bother prosecuting crimes? Well, because *of course* murderers should go to prison. As social animals we have a complex society to maintain. From a neurological perspective this is delusional, but without the delusion, the world would look a whole lot grimmer than it already is. We *have* to believe in personal responsibility and free will, or everything we've built comes crashing down. (If it really bothers you, please do be in contact and we can talk it out, so long as you acknowledge that you are not freely choosing to do so.)

The zombies lay the delusion bare—they're the only organisms that will ever be anything other than their predetermined selves.* Now, I'm not talking about you taking LSD and expanding your consciousness for a while. I'm talking about permanent, deep behavioral change. And not that this is welcome, to be clear—it's brutal and almost always ends in death for the manipulated host. So the central irony of the zombies getting to transcend the limitations of their genes is that to do so they must

---

*This to me is part of the appeal of the zombie genre. The monster is both human and not human, horrifying yet in a way an avenue for escapism. To turn zombie is to forever leave behind the bullshit of being a person on Earth—your job, your crummy apartment, any and all decisions not related to finding and consuming human flesh. Zombification is the supposed nullification of free will, which, sorry to say, never existed in the first place.

give up their bodies. Zombification beautifully demonstrates that the brain follows the laws of the physical universe: neurons arranged in a certain way firing in a certain order. Just as there is no mercurial "soul" for the manipulator to figure out how to conquer, only pure physiology, there is no soul in your head making decisions willy-nilly. The body and mind are governed by rules set from birth, which means the zombifiers have a predictable template to exploit.

Just ask Annegret Falkner. She can brain-hack her mice because she knows the password to get into the system. Simple as that. Well, it's extremely complicated neurobiology, to be sure, but it's a whole lot easier than trying to find a soul to decode and exploit. In zombifying their subjects, Falkner and the body snatchers aren't conquering some nebulous notion of free will—because it doesn't exist in the first place and they wouldn't have to anyway. By hacking material, real-world biology, they turn autonomous machines into remote-controlled playthings.

Call this the Whole-Body Model of the Great Big Sham That Is Free Will. It's tempting and perhaps excusable as humans to frame the debate over free will as a phenomenon that unfolds exclusively in the brain, because our minds happen to be so neat. But the brain is only the start. It's a governor in constant communication with both the environment and the body, a fact the manipulators have evolved to exploit so deftly. Like our friends the social parasites, which release into the environment their counterfeit pheromones, which their dupes sense with their antennae, which send the signal to the brain for processing. The parasite doesn't have to overcome any kind of free will, just a preprogrammed brain to trick by way of the environment. The ant's body is as much to blame for its downfall as its mind.

You might say, Well wait, the human brain is vastly more

complicated than that of an ant. Which is true. But all brains, no matter their complexity, follow the same physical laws of nature. So to claim that we're somehow privileged with free will while every other animal with a brain is on autopilot is an anthropocentric fallacy. It reminds me of that watchmaker argument for explaining how something as complex as the human brain could evolve. Could only a higher being have designed the convoluted human brain, and in so doing gift us and us alone with free will, so that we might "choose" a moral course in service to said higher being? Well, no—let's leave higher beings out of this. We're no more making decisions to drink tea instead of coffee than an ant decides to drink whichever it has the pleasure of stumbling upon so long as sugar is involved. Our brains just happened to evolve to trick us into thinking we maintain some sort of control over our actions.* The neurons fire, and we follow the orders.

*There*, the nihilistic dread again. Meet enough zombies and you might get to thinking that the human race is even more special than it was to you before. That these "lower" life forms get their minds snatched because they're simpletons. But by now you and I both know that's not true. The victims get their minds snatched because something figured out the code—something put a key in a lock and turned it, complexity of the brain be damned. Which means the zombifiers can conquer us, too.

---

*Why? One theory posits that it was a matter of social cohesion for our very distant ancestors. Without personal responsibility, living in small bands would have been taxing. These days you go to jail for hitting someone with a stick, which keeps a check on people hitting other people with sticks. But back then, hitting someone with a stick and saying, "Hey, my neurons told me to do it," probably wouldn't have gone over well.

# 10

## You, the Undead

*You can't sit there and tell me you think that even though you have a brain that abides by the laws of the universe, you're somehow safe from the mind hackers.*

At around ten o'clock on the evening of October 10, 1970, in the tiny Ohio town of Willshire, six-year-old Matthew Winkler awoke to find a brown bat attached to the base of his left thumb. The boy's screams summoned his father, who pried off the villain and popped it in a Mason jar, then washed the pair of puncture wounds with soap and water. A valiant effort, though in vain. The following day, the father shipped the bat to the Ohio Department of Health, which three days later reported the specimen positive for rabies. So the boy's doctor initiated a two-week course of the rabies vaccine, an invention of none other than Louis Pasteur, of pasteurization fame. Get the course going before symptoms show up, Pasteur had discovered in the 1880s, and you can stop a virus that normally kills humans without fail.

Matthew Winkler would have no such luck staving off the excruciating symptoms of rabies. Two days after finishing his inoculations, the boy complained of neck pain. Then came the dizziness, vomiting, and fever, which climbed to 104 degrees by the time he reached a hospital bed. Here he was alert enough, but grew increasingly lethargic, until a month after the bite he began having trouble walking. The virus made it difficult for the boy to talk, until it snatched his speech away entirely. "His behavior was uncooperative and bizarre," the doctors who treated him later wrote. "When efforts were made to suction saliva from his mouth, he bit and chewed on the oral suction tube." Finally, the boy descended into a coma.

But on Thanksgiving Day, 1970, Matthew managed to sit up with some assistance. He began to feed himself. Next came his first audible (albeit abnormally high-pitched, even for a six-year-old) word—"Hi." A week and a half later, he was walking with some help, and soon after his voice and cognition returned to normal. The scrappy little man who tussled with the wrong bat grew stronger and stronger, until on January 27, 1971—his seventh birthday—Matthew Winkler walked out of the hospital as the first person known to science to have survived the ravages of rabies. How, exactly, no one knows. Maybe his failed vaccines weren't total failures, and somehow helped him fight the virus. Maybe the strain the bat carried was of the low-virulence variety. Or maybe Matthew just had the gift of exceptional care at St. Rita's Hospital in Lima, Ohio, a luxury that's eluded history's countless rabies victims.

No disease is more feared and mythologized than rabies. Nothing twists its victims' bodies and minds so dramatically, turning healthy, decent people into raving psychotics so far gone that they grow terrified of drinking water. Rabies horrifies us not only because it's almost guaranteed to snatch your life from you, but because on the way there it also steals your humanity. And without our humanity, we're no different from the bats and dogs and cats that bring the virus to civilization. It's zombification, through and through, and indeed it was rabies that gave birth to the zombie we know today.

## Reality Bites

Rabies is so brutal in part because it doesn't operate like your typical virus. Instead of moving through the bloodstream, where plentiful immune cells would give it trouble, the fiend journeys through the nervous system. The problem with that is it's missing out on the bloodstream being a rapid and efficient way to move throughout a host's body. Nerve cells? Not so much. So the virus hijacks the transport systems of neurons, binding to a protein called p75, which moves freely in the cells to keep them good and healthy. (Indeed, remove p75 from nerve cells and the virus's movement slows and grows more erratic.) With the help of p75, the rabies virus marches cell by cell toward the spinal cord. Inevitably it reaches the brain, where it replicates rapidly, leading to inflammation, and moves into the salivary glands, leading to uncontrollable drooling. Thus begin the parasite's diabolical manipulations. The victim froths at the mouth, its saliva loaded with the virus. It grows hyperaggressive—even tiny creatures turn into rabid terrors, attacking animals many, many times their size.* Those victims include, of course, humans.

Like Rachel Borch. In 2017, she was jogging along a trail in Maine and came upon an unhappy raccoon that spotted her, bared its teeth, and charged. Instantly the rabid beast was at her feet. "Imagine the Tasmanian devil," she told the *Bangor Daily News*. "It was terrifying." She grabbed for the villain and tried holding it down, but instead it clamped onto her thumb. When it became clear the raccoon wasn't going to just up and let go, Borch improvised. She dragged the critter to a puddle and "pushed its head down into the muck," holding it there until "its arms sort of

---

*Another bizarre neurological pathogen called a prion produces similar symptoms of zombification in deer—drooling and aggression and lack of coordination—though it does this in a fundamentally different way. It's not a virus but a shape-shifting protein, specifically a "proteinaceous infectious particle," that attacks brain and spinal cord tissue, leading to something called chronic wasting disease. The protein folds in abnormal ways, triggering similar proteins in the host to do the same. Luckily it isn't transmissible to humans, at least not yet: Prions are also responsible for mad cow disease, which humans *can* pick up. Accordingly, the Centers for Disease Control recommends hunters not set their sights on deer acting strangely. That'd be great news for zombie deer if the disease weren't 100 percent fatal anyway.

fell to the side, its chest still heaving really slowly." Then she ran back home and made her way to the hospital to begin her series of shots. "I've never killed an animal with my bare hands," she assured the paper. "I'm a vegetarian. It was self-defense." Borch ended up surviving the beast that clearly didn't fully appreciate her dietary choices.

Such hyperviolence even manifests in rabid humans, who have been known to lash out and bite their handlers. In 2007, a rabid dog sent a twenty-eight-year-old Mumbai man to the hospital, which he proceeded to tear apart. The patient grew extremely violent, scrapping with the very people sworn to treat him. It was only with the help of cops and firefighters that the staff was able to control him, after which time the nurses took the precaution of wearing two layers of gloves and scrubs. That kept them safe enough, but the patient was hopeless. His rabies-racked body failed a week later.

Rabies is a manipulative parasite, like the wasps and worms and fungi, a scourge that transforms the behavior of its host to ensure it passes its genes to the next generation. By invading the mammalian nervous system, the virus exploits a pathway straight to the brain, twisting the mind to turn normally peaceful animals into lunatics. That makes a bite all but inevitable. The rabid victim mutates into the virus's vehicle, which crashes into still more vehicles—many more if the manipulator has sufficiently infuriated its host. Not even humans, with our fancy brains, are safe from the virus's machinations. The parasite has mastered the art of manipulating minds, tampering with the pure biology that we mammals share—nerve cell after nerve cell as it creeps toward the spinal column and into the brain. A human being is a governor, after all, like a raccoon. It was just a matter of the rabies virus cracking the code of the mammalian mind.

And rabies doesn't stop at inducing aggression in its hosts. Its most sinister manipulation is subtler: hydrophobia, literally "fear of water." Rabid animals shun water, which ensures the host doesn't wash away all that virus-packed foam accumulating in its mouth. This is of course not conducive to survival, but since rabies is almost always fatal anyway, ideally the host will have bitten new victims before perishing of dehydration.

The manipulation gets decidedly less subtle in rabid humans, though, on account of caregivers trying to force their patients to drink. The reaction is remarkable. Abandoning all self-interest, the patient recoils at the sight of a glass of water, squirming and kicking and pushing away the offering with what little strength they have left. Even if a nurse can manage to get water in the patient's mouth, excruciating spasms of the throat make swallowing nearly impossible, the virus's way of ensuring the maw is always overflowing with its weaponized foam.

I can't say whether I recommend watching videos of this online. The images are singularly horrifying, partly because you're watching a human devolve into a primal state—yet you can see the self-aware terror in their eyes as they struggle—and partly because you know full well the victim will die an excruciating death. Which is, quite frankly, better than living with the disease. Rabies is one of those things that's so terrible on so many levels, humanity can hardly bring itself to watch what it's capable of. But when we do, it's with a mixture of disbelief that we, too, can turn zombie, and a sort of compassion at seeing one of our own transformed so mercilessly. I can watch nematomorphs pile out of crickets all day, and if I had my way and the necessary permits I'd keep a colony of cockroach-hunting jewel wasps for the fun of it. But what rabies does to the human body and mind just ain't right.

A human in the throes of rabid zombification is such a powerful image, in fact, that it spawned the very phenomenon of the pop-culture zombie. The zombie of folklore first emerged in the 1600s in Haiti, where the French ran a slave trade of unimaginable brutality. The Haitians believed only death would return them to freedom in the paradise of home, of Africa—commit suicide to escape the suffering, though, and you'd be condemned to wander the island as an undead monster. Turning into a zombie, then, was to be forever imprisoned in your body, just as the enslavers tried to reduce humans into mere machines of flesh.

Several centuries later, this notion of the zombie made its way into early Hollywood in the seminal 1932 movie *White Zombie*, in which Bela Lugosi stars as the voodoo baddie Murder Legendre, perhaps the most half-assed name for a villain in all of cinema. Legendre assumes control of corpses, ordering them to rough people up and such, before messing with the wrong newlyweds and getting conked—thus releasing the spell—then proceeding to fall off a cliff. These zombies, though, weren't hungry for human flesh, and didn't carry some sort of virus to spread to their victims.

That all came in 1954, when Richard Matheson published the horror novel *I Am Legend*, in which the last normal man on Earth, Robert Neville, does battle with a human race that's turned undead. Well, technically Neville calls them vampires, but the bacterium that's spread across the planet is clearly rabies-esque.* The vampires are mindless brutes, unappreciative of garlic and

*The bacterium operates in a weirdly similar way to *Ophio*—not that anyone yet knew the complexities of the fungus's life cycle when Matheson was writing in the 1950s. When deprived of the blood that sustains it, the bacterium will "sporulate" to create an oval body. "Then, when the vampire host decomposes, these spores go flying out and seek new hosts. They find one, germinate—and one more system is infected."

sunlight and crosses and all that, but also very much the proto-zombies of modern Western pop culture—stilted, voracious, bitey—not unlike a human suffering the ravages of rabies. They're mute and curlike, Neville battling one woman who can only muster a "guttural rambling" like "the sound of a dog defending its bone." (Remember the virus snatching away little Matthew Winkler's ability to speak.) And Matheson's description of Neville tangling with an afflicted dog is particularly reminiscent of rabies: "The dog jerked around and backed into the corner, hackles rising, jaws drawn back all the way from its yellowish-white teeth, a half-mad sound quivering in its throat." The infection, of course, is transmitted by bite, which doesn't bother Neville nohow. He's immune to the vampire disease because, he reckons, a vampire bat bit him when he was stationed in Panama. (*I Am Legend* is a classic work of horror, so let's give it a pass on this plot point.)

If the whole worldwide-pandemic-that-zombifies-the-human-race shtick sounds familiar to you, it's because *I Am Legend* was a major influence for George Romero's *Night of the Living Dead,* the classic 1968 horror flick that birthed the zombie genre. A pandemic spreads, the dead rise from their graves and eat people, the still-human survivors fight them. *Night of the Living Dead* is a masterpiece that spawned the bulk of the zombie tropes we know today. Like rabies victims, the undead can't speak. They can't move too well, yet they're aggressive. They bite to spread their disease. Classic zombie stuff, and it all started with *I Am Legend* and an enduring fascination with rabies, the sole disease that strips humans of their humanity. A real-life manipulative parasite, then, made the zombies of fiction what they are today.

But rabies isn't the only disease that manipulates human behavior. Indeed, a far more common parasite may well be lurking in your brain. And for that you can thank cats.

## As If You Needed Another Reason to Get a Dog Instead of a Cat

The life of a rat is a relatively simple one: mate, avoid cats, eat, sleep, repeat. Typical governor-style behavior for an animal—until, that is, a certain parasite muscles in and makes life very confusing for the rodent.

*Toxoplasma gondii* is a protozoan, a kind of microbe that looks not unlike a shrunken comma. It can reproduce only in the gut of a cat, so it uses rats as vehicles, steering them by erasing the innate fear of their enemies. So not only do infected rats stop abhorring the scent of cat urine, they grow *attracted* to it. Cat eats rat, *Toxo* reproduces, cat poops out *Toxo*, rat eats poop, cat eats rat, etc. But the parasite doesn't rely on this mode of transmission alone—male rats also pass the microbe to females by way of their sperm.* And it turns out the parasite's two strategies are linked in a fascinating way.

When *Toxo* enters a male rat's body, it heads straight for the testes. These organs, like the brain, are immune-privileged, meaning the immune system doesn't go wild and initiate inflammation to ward off the invader. (This makes good sense in such sensitive parts of the body—the defense wouldn't want to accidentally damage sperm in the testes or neurons in the brain.) Here *Toxo* begins forming cysts, and because the testes produce testosterone, production of the hormone glitches and skyrockets.

---

*One could speculate that while a theoretical human zombie virus wouldn't need to be sexually transmissible, what with all the biting, the virus would benefit from its hosts retaining their sex drive. Indeed this is the case in the 2013 film *Warm Bodies*, in which a zombie named R falls for an uninfected woman named Julie, who wisely waits until after R has been cured before beginning a relationship.

The parasite will also encyst in the brain, but that's probably incidental—remove the testes before infection and there's no loss of fear, and remove the testes of an infected rat and the parasite's effects vanish. That and researchers have loaded rats up with a strain of *Toxo* that avoids the brain entirely, yet is still able to hijack the host's mind. So it's likely a flood of testosterone from the testes that manipulates the rodent's behavior.

Male rats, like any other animal with enemies to worry about, have to strike a balance between total aloofness and total anxiety. Hunker down in seclusion and you'll never find food or a mate, and go full-tilt out in the open and you're liable to get eaten. *Toxo* is interested in the latter. By soaking the male rat's brain in testosterone, the parasite emboldens him, destroying his fear of cats. If that lands the rat in a belly, so be it. The parasite completes its life cycle. If all that extra testosterone drives the rat into a mating frenzy, that's great, too. The parasite transmits sexually just fine. And because female rats prefer the scent of high-testosterone males, there's a good chance the infected male will get lucky. Indeed, female rats are much more likely to choose *Toxo*-infected males over healthy males. Thus you could argue that this parasite isn't entirely a burden for the male rat—in a way it's an advantage. Sure, producing more testosterone takes a lot of resources, and bounding around like a maniac takes energy. But I'll remind you that his whole purpose in life is to pass down his genes, so if a *Toxo* craze helps him mate before landing in a cat stomach, he's got an edge. Call it a performance-enhancing parasite.

But here's where things get problematic: Female rats infected with *Toxo* also lose their fear of cats, yet conspicuously lack the testes required for testosterone production on a large scale. How about the ovaries, then, if the testes are *Toxo*'s target in males? Nope—remove the ovaries and a female will still act suicidally.

Indeed, test a *Toxo*-fied female rat and you won't find any bump at all in testosterone levels. So until researchers figure out what the hell is going on here, they're operating on the theory that the parasite is exploiting the same fear-versus-aloofness paradigm in the female rat as it is in the male, perhaps by somehow tapping into the hormones that regulate her maternal instincts. That is, the rat's nurturing behaviors override her anxieties. Not that I'm saying *Toxo* makes the female rat suddenly feel like nurturing a cat—I'm saying that hormones are a powerful driver of animal behavior, as our loopy male rats make clear. And hormones are easily hacked.

Now, the problem is that beyond cats and rats, *Toxo* can infect any warm-blooded animal it chooses, including humans. If you've ever been pregnant, you've heard about the threat this parasite poses. Should a woman pick up *Toxo* just before conception or while she's carrying, the microbe can have severe impacts on the baby—including epilepsy to blindness. While the majority of human *Toxo* infections come from eating the undercooked flesh of mammals that were themselves infected, pregnant women would do well to avoid changing cat litter to be sure. (More than likely the parasite first evolved in cats, then added rats as intermediate hosts. The reason it can survive in humans is probably by virtue of us being mammals with relatively similar physiology.)

While such an infection, known as congenital toxoplasmosis, is rare, the parasite's presence in people is anything but: Perhaps one-third of humanity is infected with *Toxo*. Nearly everyone among them shows no symptoms, though if your immune system is compromised in any way, toxoplasmosis will feel like the flu. But it's what you *can't* feel that makes *Toxo* so bizarre—the microbe has been linked to schizophrenia, aggression, and even suicide risk. Which means you have a decent chance of being a parasite-manipulated zombie yourself.

## Inside the Fallible Human Brain

Schizophrenia is perhaps the most dramatic reminder that the human brain is a material thing, and the psyche it produces is necessarily grounded in neurons—physical and, unfortunately for us, corruptible cells. Schizophrenics, who make up 1 percent of the human population, battle delusions, hallucinations, fragmented speech, and extreme paranoia. Like rabies, the disease is terrifying because it strips a person of their rationality and self-preservation, oftentimes rendering them incapable of functioning in what the unafflicted would consider a normal human world. And it comes across as so bizarre because it manifests differently than your typical disease, which aggravates existing processes—a runny nose from a head cold is your body producing more mucus, and a fever from the flu is your body raising its temperature to cook off an invader. The hallucinations and delusions of schizophrenia are unique in that they have no precedent. That is, in a strange way they're a *creative* force. You and I may not hear voices, but the brain of a schizophrenic actually invents them.

And so the most incredible biological governor this planet has ever known breaks down in spectacular fashion, and no one yet knows why. As with any disease, certainly there's a genetic component that makes you more susceptible—indeed, schizophrenia runs in families. Specifically suspect is the appropriately named gene C4, which leads to "pruning" of neural connections in the developing teenage brain (which might explain why most schizophrenics don't start showing symptoms until their late teens or twenties). This pruning is usually normal, as it eliminates connections we no longer need in order to optimize the brain. But when C4 short-circuits, it may lead to runaway neural restructuring that results in schizophrenia.

Now, what I'm about to say is not that if you pick up *Toxo* you necessarily develop schizophrenia, or that the parasite is the only factor in the disease. First of all, the numbers don't compute—a third of humans carry the parasite, while 1 percent of people develop schizophrenia. That, and not all schizophrenics carry *Toxo*. And your particular genes interact with *Toxo* in unique ways, plus different strains of the parasite might have different effects. But dozens of studies have reported links between *Toxo* and schizophrenia. One review found that of fifty-four studies that explored the association, forty-nine determined schizophrenic subjects carried more *Toxo* antibodies in their blood. Yes, there are also a whole lot of genetic and environmental factors at play when it comes to this complex disease, or any disease for that matter. But researchers are building a clearer and clearer case that there's some sort of connection between *Toxo* and schizophrenia.

Consider our old friend dopamine, the same stuff the jewel wasp probably exploits to make its cockroach victims groom themselves obsessively. Like it does with rats, *Toxo* invades the human brain and forms cysts within neurons, though not with any kind of order—it doesn't target a specific part of the brain but instead grows throughout. And even then, it isn't as if the microbe takes over every neuron in the brain, as its exploits are pretty limited. But when you start messing with neurons, you start messing with neurotransmitters. Parkinson's patients, for example, suffer from the death or the crippling of neurons in the part of their brain that controls movement, leading to a dopamine crash. Schizophrenics suffer the opposite: Their dopamine levels are abnormally high. And as it happens, the same goes for humans infected with *Toxo*. Intriguingly, antipsychotic medications that treat schizophrenia lower dopamine levels, and those same medications also seem to inhibit the replication of *Toxo*. Double intriguingly, the parasite that's normally asymptomatic in the many humans it infects will sometimes lead to hallucinations—like a schizophrenic might suffer—in those with compromised immune systems. And it may be dopamine that's to blame.*

Another consideration is the timing of infection. Clearly *Toxo* is nothing but bad news for the central nervous system of a developing fetus—the epilepsy and blindness that comes with congenital toxoplasmosis. What's less clear is how infection affects the brain at various stages of adolescent growth. After all, the brain is still developing through the teenage years, and even into the midtwenties. The question then becomes: How might *Toxo* encysting in human neurons affect the development of the brain as a whole? Does it matter any if infection sets in at age five versus fifteen or twenty? Does a specific timing make it more likely that schizophrenia will develop? No one can yet say.

---

*Go ahead and also blame dopamine for your debilitating compulsion to check social media. Dopamine is a reward chemical associated with pleasures, including sex and gambling. It's a critical component in the brain from an evolutionary perspective: Making sex chemically rewarding makes an organism a good sex-seeking organism. Dopamine also spikes during social interaction, which for our long-ago ancestors might have encouraged the cohesion that was critical for their survival.

But in these times of instant social gratification on the internet, dopamine makes us chemically predisposed to get a bit carried away with it. That sort of giddiness you get when a friend likes what you've posted? That's the dopamine talking. Not that there's anything wrong with being social—but there's certainly something wrong with social media titans exploiting chemicals in your brain to make their platforms addictive. They're the parasites, and we're the hapless zombies.

Complicating matters even further, research has linked *Toxo* to subtler changes in behavior for nonschizophrenics, with men and women showing sometimes opposite effects. Infected men, for instance, tend to be more prone to jealousy and suspicion in personality tests, while infected women rate as more easygoing than their peers. (Like with male rats, this may come down to the parasite altering testosterone levels—for infected men it goes up, for women it goes down.) However, one study found that women with *Toxo* were over 50 percent more likely to commit suicide, and another found that postmenopausal women are even more at risk. And regardless of sex, infected humans tend to be more neurotic, impulsive, and aggressive in personality tests. Research has even found that sufferers of intermittent explosive disorder—which, as you might have guessed, involves sudden outbursts of often irrational anger—are twice as likely to harbor the parasite. Weirder still, four different studies have found that infected humans are more likely to end up in car accidents, which might have something to do with that aggression, or it could be the fact that in lab experiments, the *Toxo*-infected have slower reaction times.

*Toxo* has no cause to manipulate our minds. It evolved to steer rats into the stomachs of felines, where it completes its life cycle, so trying to steer humans around would be pointless. But because a rat has a mammalian brain, and you have a mammalian brain, the microbe can still make its mark. Yeah, you may think you're more "highly evolved" than a rat, but you're forgetting something very important: You're far more closely related to rodents than you are to the amphipods or cockles or ladybugs, yet here we all are, the many dupes that the many brainwashers out there use as playthings. By virtue of belonging to the vast family of living beings on Earth, we're liable to get zombified.

The problem is we're biologically fallible—all of us. These intellects we're so proud of are trapped in three pounds of squishy, hackable tissue. Every thought we think, every impulse we have, every insistence that we maintain free will, all of it is the firing of neurons. The parasites that warp our minds may not do so as extravagantly as *Ophio* commands ants, or nematomorphs turn crickets suicidal, or caterpillars brainwash whole colonies of ants, but this collective madness comes down to one universal rule: The brain is the most extraordinary thing evolution has produced in the 3.5 billion years of life on Earth, yet it remains tragically corruptible.

11

# End Times

***The zombie apocalypse is in full swing. That's both tantalizing and troubling.***

You may not want to believe me about all of this, and I can't say I blame you. I once paced around a slanted kitchen struggling with the idea of real-life zombies and also the idea of balance. You may not want to believe how vulnerable and exploitable your own brain is, or that when it comes down to it, you're not all that different from the animals you share a planet with. And you certainly don't want to believe that free will is a lie, that you're a slave to your brain and its billions of neurons, all firing out of your control.

I get that. I've been immersed in the world of the zombifiers for a long while now, and the implications that swirl around mind control haven't gotten any easier to handle. But at the risk of slipping back into my nihilistic tendencies, I must say there's something comforting about the fact that we're made of meat, just like rats and cockroaches and crustaceans, and that the brains we're so fond of follow the same laws of the physical universe. We may be the only species that can fully appreciate how parasites turn their hosts into zombies, and the only species that can conceptualize and produce a zombie movie, but we're still very much animals that run on familiar neurotransmitters like dopamine and serotonin. When you get depressed, it's not because you don't appreciate things enough. I mean, maybe you don't, but at the heart of it are your brain chemicals acting funny. Repeat after me: We are meat. We're not inadequate because we think and act weird sometimes. We are meat. You have no soul—you have only meat.

I don't want to leave you feeling rotten about all this, so let's appreciate the majesty here. Nothing Hollywood has ever concocted, or will ever concoct, can match the outlandishness of the real-life zombies that natural selection invented long before the undead first stumbled onto the silver screen. All across the tree of life, parasites have developed powers of mind control simply because they could, because for all its complexity, the brain is a predictable and therefore vulnerable machine. That means the exploiters can weaponize familiar chemicals, turning hormones against their victims, or mimic pheromones to weasel into societies where they don't belong, or devise stupefying venoms. Every bit of it is rooted in pure biology that scientists are only beginning to understand.

The secrets the zombifiers are giving up, though, are revealing a world that is diabolically different from our own—and that should tickle something in your psyche. A surgeon studies for years to manipulate the human brain in what limited capacity we're currently able to understand the organ, yet for millennia upon millennia the jewel wasp—a "lowly" insect—has been performing surprise brain surgery on cockroaches with astonishing precision. So I'm sorry to say that nothing you or I will ever do in our lives, not building a company or a house or a model train set if that's what you're into, will ever match the triumphs of the jewel wasp or any other manipulator. These are organisms that have managed to take total command of their victims.

Really, if you're going to blame anything for all this, blame sex. A mother jewel wasp never asked to spend an afternoon lobotomizing a cockroach. She and the other zombifiers unleash their shenanigans because they have to, because reproducing is their one and only mission in life. And it's this drive that made such fantastical manipulations possible in the first place. These parasites didn't show up one day with all of their seemingly supernatural powers. They evolved small step by small step, because the bits that worked helped them make offspring, and those offspring in turn passed down the bits that worked. The offspring that came up short . . . well, they didn't make it. The world is suffering, and it's this suffering that created the crustacean-steering worms and killer fungi and crafty cuckoos. Thus from destruction come manipulations so creative, they seem impossible.

Yet zombification is so common, having evolved so many times across the tree of life, that there's also something fundamentally *un*creative about it: Parasitic manipulation is such an effective strategy that unrelated organisms have stumbled upon startlingly similar strategies. An acanthocephalan can short-circuit an amphipod's mind by getting it to release a flood of serotonin, for instance, making the dupe mistake light for darkness, while an unrelated trematode has evolved the same way to manipulate the same host. And remember the worms that invade cockles and seize up their feet to expose them to predators, while in reef ecosystems another worm invades coral polyps, seizing them up and exposing them to their own predators. And consider the killer caterpillars that brainwash whole colonies of ants with propaganda pheromones, before the wasp arrives and releases its own pheromones to glitch the colony. Zombification isn't just common—it's common to the point where the zombifiers are stealing each other's ideas.

This has been a book of conflict, and necessarily so—you're not making friends if you're a fungus growing through an ant's muscles and wrenching apart the fibers. But the manipulators remind us better than anything else in nature that every species is connected, the collective ancestors of one ancient organism that started it all. No matter how different we may seem, no matter how many legs we run on or fins we flap, we share the common language of DNA. We humans rely on dopamine, just as the bugs do. The social parasites copy their hosts' pheromones to infiltrate the colony. The rabies virus follows a map through the nervous system and into the mammalian brain. As alien as it may all seem, there's a perfectly rational explanation for it: good old fallible biology.

That we know this much is astonishing, but there remains so much more that we may well never know. That frustrates me a bit, but there's also beauty in not knowing. After all, we're stuck in this umwelt, hardly capable of sensing outside the range of sights and smells and sounds our bodies have evolved to detect. So it's an impossibility that the world of the body snatchers isn't much bigger than we realize. The weirdest, subtlest, most intricate manipulations still await discovery, and that's exhilarating and humbling, if you ask me. Really, we're lucky to be the only animals on Earth that can even begin to escape our umwelt—thanks, at least in part, to a little technological marvel called the Pitch Computer PC 1400.

Well, you gotta start somewhere.

# Acknowledgments

Every so often you find yourself in a New Mexico forest hunting for kamikaze crickets with a man named Ben Hanelt, who's very reassuring about bears and cougars. And you'll want to thank him. So thanks, Ben. And thank you David Hughes, for finally introducing me to my first zombie ant. It was an honor. Thanks to Annegret Falkner for teaching me how to whisper to mice by way of neurosurgery. And Frederic Libersat: much obliged to you and your wasps, those beloved torturers of cockroaches.

A no-less-emphatic thank you to the people I never shook hands with, yet who were instrumental in helping me write this book. Shelley Adamo, a founding member of the field of parasitic manipulations and a wonderfully enthusiastic scientific personality. Same goes for Janice Moore, the parasitology pioneer who helped get me out of my umwelt. Marie-Jeanne Perrot-Minnot and her appreciation of the underappreciated worms. Robert Poulin and those cockles you were so brave to consume. Kim Fleming for introducing me to uniquely American zombie ants. Tommy Leung for excellent conversation about the kingdom of the parasites. And here's to the three dons of *Toxo*—Glenn McConkey and Ajai Vyas and Jaroslav Flegr—for your insight into a manipulative parasite I'd rather not think about, if I'm being honest.

I can't thank enough the amazing people who were so kind as to peep the manuscript. I'm in your debt. Danielle Venton, may your days be populated with beer and ungulates. Daniela Hernandez, brain person and brainy person. Gwen Pearson, biologist extraordinaire and insect evangelist—need I say more. Samantha Oltman, keep on skatin', or whatever it is the kids say. Nick Stockton, an uplifting editor and all-around fantastic fellow, even though he likes cats. Ben Johnston, who looked at the book even though he knew full well the decision wasn't his to make.

Thank you to my agent and all around person-with-way-more-energy-for-this-hustle-than-I-have, David Fugate. And to my editor Meg Leder, and to Shannon Kelly and the whole incredible crew at Penguin—you make this so, so easy and so, so fun.

Lauren, for your positivity and brilliance and wildly-irrational-by-this-point support. This book, however, is *not* for the creature who made it nearly impossible: Cricket the cat. You're an ingrate, and I don't like you.

Lastly, this book is for the people who made it possible, in the sense that they are responsible for bringing me into existence: Mom and Dad and Melissa and Grandma, I love ya. Grandpa, thanks for my weird outlook on the world. You almost got me to not hate golfing.

# Bibliography

## 1
## The First Rule of Zombification: You Do Not Want to Be a Zombie

Dheilly, N., et al. (2015). Who Is the Puppet Master? Replication of a Parasitic Wasp-Associated Virus Correlates with Host Behaviour Manipulation. *Proceedings of the Royal Society B*, Vol. 282, Issue 1803.

Emanuel, S., and Libersat, F. (2017). Do Quiescence and Wasp Venom-Induced Lethargy Share Common Neuronal Mechanisms in Cockroaches? *PLoS ONE*, Vol. 12, Issue 1.

Gal, R., et al. (2014). Sensory Arsenal on the Stinger of the Parasitoid Jewel Wasp and Its Possible Role in Identifying Cockroach Brains. *PLoS ONE*, Vol. 9, Issue 2.

Gal, R., and Libersat, F. (2010). A Wasp Manipulates Neuronal Activity in the Sub-Esophageal Ganglion to Decrease the Drive for Walking in Its Cockroach Prey. *PLoS ONE*, Vol. 5, Issue 4.

Gavra, T., and Libersat, F. (2010). Involvement of the Opioid System in the Hypokinetic State Induced in Cockroaches by a Parasitoid Wasp. *Journal of Comparative Physiology*, Vol. 197, Issue 3.

Haspel, G., et al. (2005). Parasitoid Wasp Affects Metabolism of Cockroach Host to Favor Food Preservation for Its Offspring. *Journal of Comparative Physiology*, Vol. 191, Issue 6.

Haspel, G., Rosenberg, L., and Libersat, R. (2003). Direct Injection of Venom by a Predatory Wasp into Cockroach Brain. *Developmental Neurobiology*, Vol. 56, Issue 3.

Libersat, F. (2003). Wasp Uses Venom Cocktail to Manipulate the Behavior of Its Cockroach Prey. *Journal of Comparative Physiology*, Vol. 189, Issue 7.

Libersat, F. (2009). Neuroethology of Parasitoid Wasps. *Scholarpedia*, Vol. 4, Issue 7.

Libersat, F., and Gal, R. (2007). Neuro-Manipulation of Hosts by Parasitoid Wasps. *Recent Advances in the Biochemistry, Toxicity, and Mode of Action of Parasitic Wasp Venoms*.

Libersat, F., and Gal, R. (2013). What Can Parasitoid Wasps Teach Us About Decision-Making in Insects? *Journal of Experimental Biology*, Vol. 216.

Libersat, F., and Gal, R. (2014). Wasp Voodoo Rituals, Venom-Cocktails, and the Zombification of Cockroach Hosts. *Integrative and Comparative Biology*, Vol. 54, No. 2.

Maure, F., et al. (2011). The Cost of a Bodyguard. *Royal Society Biology Letters*, 10.1098/rsbl.2011.0415.

Ohl, M. (2014). The Soul-Sucking Wasp by Popular Acclaim—Museum Visitor Participation in Biodiversity Discovery and Taxonomy. *PLoS ONE*, Vol. 9, Issue 4.

Simon, M. (2015). Absurd Creature of the Week: If This Wasp Stings You, 'Just Lie Down and Start Screaming.' *Wired*. Retrieved from https://www.wired.com/2015/07/absurd-creature-of-the-week-tarantula-hawk/.

Takasuka, K., et al. (2015). Host Manipulation by an Ichneumonid Spider Ectoparasitoid That Takes Advantage of Preprogrammed Web-Building Behaviour for Its Cocoon Protection. *Journal of Experimental Biology*, Vol. 218.

Weiler, N. (2015). Wasp Virus Turns Ladybugs into Zombie Babysitters. *Science*. Retrieved from http://www.sciencemag.org/news/2015/02/wasp-virus-turns-ladybugs-zombie-babysitters.

Wulf, A. (2015). *The Invention of Nature: The Adventures of Alexander Von Humboldt*. London: John Murray.

## 2
## Nothing Brings the World Together Like Unsolicited Mind Control

Blume, M. (2008). France's Unsolved Mystery of the Poisoned Bread. *New York Times*. Retrieved from http://www.nytimes.com/2008/07/24/arts/24iht-blume.1.14718462.html.

Darwin, C. (1858). Letter to Charles Lyell. https://www.darwinproject.ac.uk/letter/DCP-LETT-2285.xml.

Darwin, C. (1911). *The Life and Letters of Charles Darwin: Including an Autobiographical Chapter, Volume 2*. New York and London: D. Appleton.

De Bekker, C., et al. (2014). Species-Specific Ant Brain Manipulation by a Specialized Fungal Parasite. *BMC Evolutionary Biology*, 10.1186/s12862-014-0166-3.

De Bekker, C., et al. (2015). Gene Expression During Zombie Ant Biting Behavior Reflects the Complexity Underlying Fungal Parasitic Behavioral Manipulation. *BMC Genomics*, 10.1186/s12864-015-1812-x.

The Emperor's Mighty Brother (2015). *Economist*. Dec. 19. Retrieved from http://www.economist.com/news/christmas-specials/21683980-demand-aphrodisiac-has-brought-unprecedented-wealth-rural-tibetand-trouble.

Evans, H., Elliot, S., and Hughes, D. (2011). Hidden Diversity Behind the Zombie-Ant Fungus *Ophiocordyceps unilateralis*: Four New Species Described from Carpenter Ants in Minas Gerais, Brazil. *PLoS ONE* 6(3): e17024.10.1371/journal.pone.0017024.

Finkel, M. (2012). Tibet's Golden "Worm." *National Geographic*. Retrieved from http://ngm.nationalgeographic.com/2012/08/tibetan-mushroom/finkel-text.

Gabbai, Lisbonne, and Pourquier (1951). Ergot Poisoning at Pont St. Esprit. *British Medical Journal* 2(4732): 650–651.

Hoffman, A. (2013). *LSD: My Problem Child*. Oxford: Oxford University Press.

Hughes, D., et al. (2011). Behavioral Mechanisms and Morphological Symptoms of Zombie Ants Dying from Fungal Infection. *BMC Ecology* 11:13.

Krasnoff, S., Watson, D., Gibson, D., and Kwan, E. (1995). Behavioral Effects of the Entomopathogenic Fungus, *Entomophthora muscae* on Its Host *Musca domestica*: Postural Changes in Dying Hosts and Gated Pattern of Mortality. *Journal of Insect Physiology*, Vol. 41, No. 10, 895–903.

Labaude, S., Rigaud, T., and Cézilly, F. (2015). Host Manipulation in the Face of Environmental Changes: Ecological Consequences. *International Journal for Parasitology: Parasites and Wildlife*, Vol. 4, Issue 3.

Lehmann, T. (2013). Outracing All Your Devoted Enemies? The Periodic Cicada (and Its Bizarre Fungal Pathogen). *Fungi*, Vol. 6, No. 3.

Lev-Yadun, S. (2016). *Defensive (Anti-Herbivory) Coloration in Land Plants*. Switzerland: Springer.

Lo, H., Hsieh, C., Lin, F., and Hsu, T. (2013). A Systematic Review of the Mysterious Caterpillar Fungus *Ophiocordyceps sinensis* in Dong-ChongXiaCao (冬蟲夏草 Dōng Chóng Xià Cǎo) and Related Bioactive Ingredients. *Journal of Traditional and Complementary Medicine* 3(1): 16–32.

Macias, A. (2013). Flying Salt Shakers of Death. Cornell Mushroom Blog. Retrieved from http://blog.mycology.cornell.edu/2013/02/19/flying-salt-shakers-of-death/.

Maitland, D. (1994). A Parasitic Fungus Infecting Yellow Dungflies Manipulates Host Perching Behaviour. *Proceedings: Biological Sciences*, Vol. 258, No. 1352.

McCalman, I. (2009). *Darwin's Armada: Four Voyages and the Battle for the Theory of Evolution*. New York: W. W. Norton.

Rhodes, J. (2011). How Deadly Bread Bewitched a French Village. *Smithsonian*. Retrieved from http://www.smithsonianmag.com/arts-culture/how-deadly-bread-bewitched-a-french-village-123126177/.

Speare, A. (1921). Massospora cicadina Peck: A Fungous Parasite of the Periodical Cicada. *Mycologia*, Vol. 13, No. 2.

Stone, R. (2008). Last Stand for the Body Snatcher of the Himalayas? *Science*, Vol. 322, Issue 5905.

Wallace, A. (1858). Letter to Joseph Hooker. Retrieved from https://www.darwinproject.ac.uk/letter/DCP-LETT-2337.xml.

Wallace, A. (1905). *My Life: A Record of Events and Opinions*. London: Chapman and Hall.

Wellcome Library (1958). Ergot: The Story of a Parasitic Fungus. Retrieved from https://www.youtube.com/watch?v=ielb0C1JYsw.

Zuk, M., Thornhill, R., Ligon, J., and Johnson, K. (1990). Parasites and Mate Choice in Red Jungle Fowl. *American Zoologist*, Vol. 30, No. 2.

Zurek, L., Watson, D., Krasnoff, S., and Schal, C. (2002). Effect of the Entomopathogenic Fungus, *Entomophthora muscae* (Zygomycetes: Entomophthoraceae), on Sex Pheromone and Other Cuticular Hydrocarbons of the House Fly, *Musca domestica*. *Journal of Invertebrate Pathology* 80, 171–176.

## 3
## When Life Gets Complicated, Life Gets Zombified

Bakker, T., Mazzi, D., and Zala, S. (1997). Parasite-Induced Changes in Behavior and Color Make *Gammarus pulex* More Prone to Fish Predation. *Ecology*, Vol. 78, No. 4.

Baldauf, S., et al. (2007). Infection with an Acanthocephalan Manipulates an Amphipod's Reaction to a Fish Predator's Odours. *International Journal for Parasitology*, Vol. 37.

Beebe, W. (1941). *Edge of the Jungle*. New York: Garden City Publishing.

Benesh, D., et al. (2009). Seasonal Changes in Host Phenotype Manipulation by an Acanthocephalan: Time to Be Transmitted? *Parasitology*, Vol. 136, Issue 2.

Bethel, W., and Holmes, J. (1973). Altered Evasive Behavior and Responses to Light in Amphipods Harboring Acanthocephalan Cystacanths. *Journal of Parasitology*, Vol. 59, No. 6.

Carney, W. (1969). Behavioral and Morphological Changes in Carpenter Ants Harboring *Dicrocoeliid metacercariae*. *American Midland Naturalist*, Vol. 82, No. 2.

Cornell Lab of Ornithology (2008). The Basics: Feather Molt. Retrieved from https://www.allaboutbirds.org/the-basics-feather-molt/.

Delsuc, F. (2003). Army Ants Trapped by Their Evolutionary History. *PLoS Biology*, 10.1371/journal.pbio.0000037.

Helluy, S. (2013). Parasite-Induced Alterations of Sensorimotor Pathways in Gammarids: Collateral Damage of Neuroinflammation? *Journal of Experimental Biology*, Vol. 216.

Leung, T. (2016). *Confluaria podicipina*. Parasite of the Day. Retrieved from http://dailyparasite.blogspot.com/2016/03/confluaria-podicipina.html.

Leung, T. (2016). *Tylodelphys* sp. Parasite of the Day. Retrieved from http://dailyparasite.blogspot.com/2016/09/tylodelphys-sp.html.

Medoc, V., et al. (2009). A Manipulative Parasite Increasing an Antipredator Response Decreases Its Vulnerability to a Nonhost Predator. *Animal Behavior*, Vol. 77.

Medoc, V., and Beisal, J. (2008). An Acanthocephalan Parasite Boosts the Escape Performance of Its Intermediate Host Facing Non-Host Predators. *Parasitology*, Vol. 135.

Paley, W. (1802). *Natural Theology: or, Evidences of the Existence and Attributes of the Deity, Collected from the Appearances of Nature.* London: Wilks and Taylor.

Pérez-López, R., et al. (2011). Evaluation of Heavy Metals and Arsenic Speciation Discharged by the Industrial Activity on the Tinto-Odiel Estuary, SW Spain. *Marine Pollution Bulletin*, Vol. 62, Issue 2.

Perrot-Minnot, M., Sanchez-Thirion, K., and Cézilly, F. (2014). Multidimensionality in Host Manipulation Mimicked by Serotonin Injection. *Proceedings of the Royal Society B*, 10.1098/rspb.2014.1915.

Poinar, G., and Yanoviak, S. (2008). *Myrmeconema neotropicum* n. g., n. sp., a New Tetradonematid Nematode Parasitising South American Populations of *Cephalotes atratus* (Hymenoptera: Formicidae), with the Discovery of an Apparent Parasite-Induced Host Morph. *Systematic Parasitology*, Vol. 69, Issue 2.

Ponton, F., et al. (2005). Ecology of Parasitically Modified Populations: A Case Study from a Gammarid-Trematode System. *Marine Ecology Progress Series*, Vol. 299.

Poulin, R. (2007). *Evolutionary Ecology of Parasites*. Princeton, NJ: Princeton University Press.

Sánchez, M., et al. (2009). Neurological and Physiological Disorders in Artemia Harboring Manipulative Cestodes. *The Journal of Parasitology*, Vol. 95, No. 1.

Sánchez, M., et al. (2016). When Parasites Are Good for Health: Cestode Parasitism Increases Resistance to Arsenic in Brine Shrimps. *PLoS Pathogens*, Vol. 12, No. 3.

Sanders, R. (2008). Ant Parasite Turns Host into Ripe Red Berry, Biologists Discover. *UC Berkeley News*. Retrieved from http://www.berkeley.edu/news/media/releases/2008/01/16_ants.shtml.

Simon, M. (2014). Absurd Creature of the Week: The Parasitic Worm That Turns Snails into Disco Zombies. *Wired*. https://www.wired.com/2014/09/absurd-creature-of-the-week-disco-worm/.

Simon, M. (2014). Absurd Creature of the Week: World's Most Badass Ant Skydives, Uses Own Head as a Shield. *Wired*. Retrieved from https://www.wired.com/2014/04/absurd-creature-of-the-week-the-amazing-skydiving-ant/.

Stumbo, A., and Poulin, R. (2016). Possible Mechanism of Host Manipulation Resulting from a Diel Behaviour Pattern of Eye-Dwelling Parasites? *Parasitology*, Vol. 143.

Tain, L., Perrot-Minnot, M., and Cézilly, F. (2006). Altered Host Behaviour and Brain Serotonergic Activity Caused by Acanthocephalans: Evidence for Specificity. *Proceedings of the Royal Society B*, Vol. 273.

Wesołowska, W., and Wesołowska, T. (2013). Do Leucochloridium Sporocysts Manipulate the Behaviour of Their Snail Hosts? *Journal of Zoology*, Vol. 292, Issue 3.

Wickler, W. (1976). Evolution-Oriented Ethology, Kin Selection, and Altruistic Parasites. *Zeitschrift für Tierpsychologie*, Vol. 42, Issue 2.

## 4
## No Creature Lives in a Vacuum, Not Even a Zombie

Aeby, G. (1991). Behavioral and Ecological Relationships of a Parasite and Its Hosts Within a Coral Reef System. *Pacific Science*, Vol. 45, No. 3.

Arkive.org (n.d.). Stony Coral (*Porites compressa*). Retrieved from http://www.arkive.org/stony-coral/porites-compressa/.

Babirat, C., Mouritsen, K., and Poulin, R. (2004). Equal Partnership: Two Trematode Species, Not One, Manipulate the Burrowing Behaviour of the New Zealand Cockle, *Austrovenus stutchburyi*. *Journal of Helminthology* 78(3): 195–199.

Capinera, J. (2000). Featured Creatures: Imported Cabbageworm. University of Florida. Retrieved from http://entnemdept.ufl.edu/creatures/veg/leaf/imported_cabbageworm.htm.

Cézilly, F., Gregiore, A., and Bertin, A. (2000). Conflict Between Co-Occurring Parasites? An Experimental Study of the Joint Influence of Two Acanthocephalan Parasites on the Behaviour of *Gammarus pulex*. *Parasitology* 120(6): 625–630.

Chrisler, L. (1956). Observations of Wolves Hunting Caribou. *Journal of Mammalogy*, Vol. 37, No. 3.

Dezfuli, B., Giari, L., and Poulin, R. (2000). Species Associations Among Larval Helminths in an Amphipod Intermediate Host. *International Journal for Parasitology*, Vol. 30.

Haine, E., Boucansaud, K., and Rigaud, T. (2005). Conflict Between Parasites with Different Transmission Strategies Infecting an Amphipod Host. *Proceedings of the Royal Society B*, 272, 2505–2510.

Hernandez, A., and Sukhdeo, M. (2008). Parasite Effects on Isopod Feeding Rates Can Alter the Host's Functional Role in a Natural Stream Ecosystem. *International Journal for Parasitology*, Vol. 38.

Joly, D., and Messier, F. (2004). The Distribution of *Echinococcus granulosus* in Moose: Evidence for Parasite-Induced Vulnerability to Predation by Wolves? *Oecologia*, Vol. 140, No. 4.

Lafferty, K. (1999). The Evolution of Trophic Transmission. *Parasitology Today* 15:111–115.

Lafferty, K., and Morris, A. (1996). Altered Behavior of Parasitized Killifish Increases Susceptibility to Predation by Bird Final Hosts. *Ecology*, Vol. 77, No. 5.

Lefèvre, T., et al. (2008). The Ecological Significance of Manipulative Parasites. *Trends in Ecology and Evolution*, Vol. 24, Issue 1.

Mahr, S. (n.d.). *Cotesia glomerata*, Parasite of Imported Cabbageworm. University of Wisconsin. Retrieved from http://www.entomology.wisc.edu/mbcn/kyf303.html.

Młot, C. (2016). Extreme Inbreeding Likely Spells Doom for Isle Royale Wolves. *Science*. Retrieved from http://www.sciencemag.org/news/2016/04/extreme-inbreeding-likely-spells-doom-isle-royale-wolves.

Mouritsen, K., and Poulin, R. (2003). Parasite-Induced Trophic Facilitation Exploited by a Non-Host Predator: A Manipulator's Nightmare. *International Journal for Parasitology*, Vol. 33, Issue 10.

Mouritsen, K., and Poulin, R. (2005). Parasites Boost Biodiversity and Change Animal Community Structure by Trait-Mediated Indirect Effects. *Oikos*, Vol. 108, No. 2.

Mouritsen, K., and Poulin, R. (2010). Parasitism as a Determinant of Community Structure on Intertidal Flats. *Marine Biology*, Vol. 157, Issue 1.

Peterson, R. (2007). *The Wolves of Isle Royale: A Broken Balance*. Ann Arbor: University of Michigan Press.

Shelton, A. (n.d.). *Cotesia glomerata*. A Guide to Natural Enemies in North America. Retrieved from https://biocontrol.entomology.cornell.edu/parasitoids/cotesia.php.

Simon, M. (2017). Mathematicians Decode the Surprising Complexity of Cow Herds. *Wired*. Retrieved from https://www.wired.com/story/cow-herds/.

Thomas, F., et al. (1997). Hitch-Hiker Parasites or How to Benefit from the Strategy of Another Parasite. *Evolution*, 10.2307/2411060.

Thomas, F., et al. (1998). Manipulation of Host Behaviour by Parasites: Ecosystem Engineering in the Intertidal Zone? *Royal Society Proceedings B*, 265(1401).

Thomas, F., Renaud, F., and Poulin, R. (1998). Exploitation of Manipulators: "Hitch-Hiking" as a Parasite Transmission Strategy. *Animal Behaviour* 56(1): 199–206.

Wolchover, N. (2011). Can a Butterfly in Brazil Really Cause a Tornado in Texas? *Live Science*. Retrieved from https://www.livescience.com/17455-butterfly-effect-weather-prediction.html.

## 5
## How to Succeed in Parasitism Without Really Dying

Anderson, R., Blanford, S., Jenkins, N., and Thomas, M. (2013). Discriminating Fever Behavior in House Flies. *PLoS ONE*, 8(4): e62269.

Anderson, R., Blanford, S., and Thomas, M. (2013). House Flies Delay Fungal Infection by Fevering: at a Cost. *Ecological Entomology*, 10.1111/j.1365-2311.2012.01394.x.

Combes, C. (2005). *The Art of Being a Parasite*. Chicago: University of Chicago Press.

Ernst, C., Hanelt, B., and Buddle, C. (2016). Parasitism of Ground Beetles (Coleoptera: Carabidae) by a New Species of Hairworm (Nematomorpha: Gordiida) in Arctic Canada. *Journal of Parasitology* 102(3): 327–335.

Franceschi, N., et al. (2010). Co-Variation Between the Intensity of Behavioural Manipulation and Parasite Development Time in an Acanthocephalan–Amphipod System. *Journal of Evolutionary Biology*, 10.1111/j.1420-9101.2010.02076.x.

Libersat, F., and Moore, J. (2000). The Parasite *Moniliformis moniliformis* Alters the Escape Response of Its Cockroach Host *Periplaneta americana*. *Journal of Insect Behavior*, Vol. 13, No. 1.

Macnab, V., and Barber, I. (2011). Some (Worms) Like It Hot: Fish Parasites Grow Faster in Warmer Water, and Alter Host Thermal Preferences. *Global Change Biology*, Vol. 18, Issue 5.

Maure, F., et al. (2011). The Cost of a Bodyguard. *Royal Society Biology Letters*, 10.1098/rsbl.2011.0415.

Maure, F., Brodeur, J., Hughes, D., and Thomas, F. (2013). How Much Energy Should Manipulative Parasites Leave to Their Hosts to Ensure Altered Behaviours? *Journal of Experimental Biology* 216:43–46, 10.1242/jeb.073163.

Moore, J. (1983). Responses of an Avian Predator and Its Isopod Prey to an Acanthocephalan Parasite. *Ecology*, Vol. 64, No. 5.

Moore, J., Freehling, M., and Gotelli, N. (1994). Altered Behavior in Two Species of Blattid Cockroaches Infected with *Moniliformis*. *Journal of Parasitology*, Vol. 80, No. 2.

Muller, C., and Schmid-Hempel, P. (1993). Exploitation of Cold Temperature as Defense Against Parasitoids in Bumblebees. *Nature*, 10.1038/363065a0.

Natural History Museum of Los Angeles County (2013). The Conopid's Clutch. Retrieved from http://www.nhm.org/nature/blog/conopids-clutch.

Sato, T., Arizono, M., Sone, R., and Harada, Y. (2008). Parasite-Mediated Allochthonous Input: Do Hairworms Enhance Subsidized Predation of Stream Salmonids on Crickets? *Canadian Journal of Zoology* 86(3): 231-235, 10.1139/Z07-135.

Watson, D., Mullens, B., and Peterson, J. (1993). Behavioral Fever Response of *Musca domestica* (Diptera: Muscidae) to Infection by *Entomophthora muscae* (Zygomycetes: Entomophthorales). *Journal of Invertebrate Pathology*, 10.1006/jipa.1993.1003.

Yong, E. (2014). Parasite Forces Host to Dig Its Own Grave. *National Geographic*. Retrieved from http://phenomena.nationalgeographic.com/2014/05/20/parasite-forces-host-to-dig-its-own-grave/.

## 6
## Dawn of the Sexually Undead

Annunziato, A. (2008). DNA Packaging: Nucleosomes and Chromatin. *Nature Education*. Retrieved from http://www.nature.com/scitable/topicpage/dna-packaging-nucleosomes-and-chromatin-310.

Beani, L. (2006). Crazy Wasps: When Parasites Manipulate the *Polistes* Phenotype. *Annales Zoologici Fennici*, Vol. 43, No. 5/6.

Beani, L., et al. (2011). When a Parasite Breaks All the Rules of a Colony: Morphology and Fate of Wasps Infected by a Strepsipteran Endoparasite. *Animal Behaviour*, Vol. 82.

Centers for Disease Control and Prevention (n.d.). Parasites—Schistosomiasis. Retrieved from https://www.cdc.gov/parasites/schistosomiasis/.

Chang, S. (2016). A California City Is Fending Off Zika by Releasing 40,000 Mosquitoes Every Week. *Wired*. Retrieved from https://www.wired.com/2016/08/california-city-fending-off-zika-releasing-40000-mosquitoes-every-week/.

Darwin, C. (1854). *A Monograph on the Sub-Class Cirripedia, with Figures of All the Species*. London: Ray Society.

Darwin, C. (2000). *Charles Darwin's Zoology Notes and Specimen Lists from H. M. S.* Beagle. Cambridge: Cambridge University Press.

Darwin Correspondence Project. Darwin in Letters, 1847–1850: Microscopes and Barnacles. University of Cambridge. Retrieved from https://www.darwinproject.ac.uk/letters/darwins-life-letters/darwin-letters-1847-1850-microscopes-and-barnacles.

De Crespigny, F. E., Pitt, T. D., and Wednell, N. (2006). Increased Male Mating Rate in Drosophila Is Associated with *Wolbachia* Infection. *Journal of Evolutionary Biology*, Vol. 19, Issue 6.

Høeg, J. (1995). The Biology and Life Cycle of the Rhizocephala (Cirripedia). *Journal of the Marine Biological Association of the UK*, Vol. 75, Issue 3.

Hooker, J. (1845). Letter No. 914. Darwin Correspondence Project. Retrieved from https://www.darwinproject.ac.uk/letter/DCP-LETT-914.xml.

Hughes, D. (2005). Parasitic Manipulation: A Social Context. *Behavioural Processes*, Vol. 68.

Hughes, D., et al. (2004). Social Wasps Desert the Colony and Aggregate Outside if Parasitized: Parasite Manipulation? *Behavioral Ecology*, Vol. 15, No. 6.

Koukou, K. (2006). Influence of Antibiotic Treatment and *Wolbachia* Curing on Sexual Isolation Among *Drosophila melanogaster* Cage Populations. *Evolution*, Vol. 60, No. 1.

Lafferty, K., and Kuris, A. (2009). Parasitic Castration: The Evolution and Ecology of Body Snatchers. *Trends in Parasitology*, Vol. 25, No. 12.

Leung, T. (2014). The Crab-Castrating Parasite That Zombifies Its Prey. *Conversation*. Retrieved from http://theconversation.com/the-crab-castrating-parasite-that-zombifies-its-prey-27200.

Minchella, D., and Loverde, P. (1981). A Cost of Increased Early Reproductive Effort in the Snail *Biomphalaria glabrata*. *American Naturalist*, Vol. 118, No. 6.

National Institutes of Health (2016). Schizophrenia's Strongest Known Genetic Risk Deconstructed. Retrieved from https://www.nih.gov/news-events/news-releases/schizophrenias-strongest-known-genetic-risk-deconstructed.

Riparbelli, M., et al. (2012). *Wolbachia*-Mediated Male Killing Is Associated with Defective Chromatin Remodeling. *PLoS ONE* 7(1): e30045.

Simon, M. (2015). Absurd Creature of the Week: The Barnacle That Invades Crabs in a Not OK Way. *Wired*. Retrieved from https://www.wired.com/2015/07/absurd-creature-of-the-week-rhizocephalan/.

Sorensen, R., and Minchella, D. (2001). Snail-Trematode Life History Interactions: Past Trends and Future Directions. *Parasitology*, Vol. 123.

Stamos, D. (2007). *Darwin and the Nature of Species*. Albany: State University of New York Press.

Stott, R. (2003). How Mr Arthrobalanus Saved Charles Darwin from Baron Munchausen's Fate. *Times Higher Education*. Retrieved from https://www.timeshighereducation.com/features/how-mr-arthrobalanus-saved-charles-darwin-from-baron-munchausens-fate/175339.article.

Toscano, B., Newsome, B., and Griffen, B. (2014). Parasite Modification of Predator Functional Response. *Oecologia*, Vol. 175, Issue 1.

Vandekerckhove, T. (2003). Evolutionary Trends in Feminization and Intersexuality in Woodlice (Crustacea, Isopoda) Infected with *Wolbachia pipientis* (α-Proteobacteria). *Belgian Journal of Zoology*, Vol. 133, Issue 1.

Werren, J., Baldo, L., and Clark, M. (2008). *Wolbachia*: Master Manipulators of Invertebrate Biology. *Nature Reviews*, Vol. 6.

World Health Organization (n.d.). Schistosomiasis. Retrieved from http://www.who.int/mediacentre/factsheets/fs115/en/.

Zelnio, K. (2010). Ex Omnia Conchis: Darwin and His Beloved Barnacles. *Deep Sea News*. Retrieved from http://www.deepseanews.com/2010/02/ex-omnia-conchis-darwin-and-his-beloved-barnacles/.

## 7
## The Great Escape from the Umwelt

Brentar, C. (2015). *Jakob von Uexküll: The Discovery of the Umwelt Between Biosemiotics*. New York: Springer.

Brulliard, K. (2016). This Man Lived as a Goat for Nearly a Week. We Asked Him Why. [How's *that* for a headline.] *Washington Post*. Retrieved from https://www.washingtonpost.com/news/animalia/wp/2016/05/25/this-man-lived-as-a-goat-for-nearly-a-week-we-asked-him-why/.

Buchanan, B. (2008). *Onto-Ethologies: The Animal Environments of Uexkull, Heidegger, Merleau-Ponty, and Deleuze*. Albany: State University of New York.

Clench, M., and Mathias, J. (1995). The Avian Cecum: A Review. *Wilson Bulletin*, Vol. 107, No. 1.

Cox, R. (2004). Population Ecology of the Red Grouse, *Lagopus lagopus scoticus*, with Particular Reference to the Effects of the Parasite *Trichostrongylus tenuis*. Durham theses, Durham University.

Cressey, D. (2012). Pigeons May "Hear" Magnetic Fields. *Nature*. Retrieved from http://www.nature.com/news/pigeons-may-hear-magnetic-fields-1.10540.

Foelix, R., and Axtell, R. (1972). Ultrastructure of Haller's Organ in the Tick *Amblyomma americanum*. *Cell and Tissue Research*, Volume 124, Issue 3.

Galambos, R. (1942). The Avoidance of Obstacles by Flying Bats: Spallanzani's Ideas (1794) and Later Theories. *Isis*, Vol. 34, No. 2.

Goldsmith, T. (2006). What Birds See. *Scientific American*, June, 68–75.
Hanson, J. (2013). How Bees Can See the Invisible. It's Okay to Be Smart. Retrieved from https://www.youtube.com/watch?v=N1TUDFCOwjY.
Hudson, P., Dobson, A., and Newborn, D. (1992). Do Parasites Make Prey Vulnerable to Predation? Red Grouse and Parasites. *Journal of Animal Ecology*, Vol. 61, No. 3.
Liang, C. H., et. al. (2016). Magnetic Sensing Through the Abdomen of the Honey Bee. *Nature*. 10.1038/srep23657.
McGann, J. (2017). Poor Human Olfaction Is a 19th-Century Myth. *Science*, Vol. 356, Issue 6338.
Moore, J. (2013). An Overview of Parasite-Induced Behavioral Alterations—and Some Lessons from Bats. *Journal of Experimental Biology*, Vol. 216.
Mullen, G., and Durden, L. (2009). *Medical and Veterinary Entomology*. London: Elsevier.
Obenchain, F., and Galun, R. (1982). *Physiology of Ticks: Current Themes in Tropical Science*. London: Elsevier.
Rempel, J. (1940). Intersexuality in Chironomidae Induced by Nematode Parasitism. *Journal of Experimental Zoology*, Vol. 84, Issue 2.
Sonenshine, D., Taylor, D., and Carson, K. (1986). Chemically Mediated Behavior in Acari: Adaptations for Finding Hosts and Mates. *Journal of Chemical Ecology*, Vol. 12, Issue 5.
Von Uexküll, J. (1926). *Theoretical Biology*. New York: Harcourt, Brace.
Von Uexküll, J. (2010). *A Foray into the Worlds of Animals and Humans: With a Theory of Meaning*. Minneapolis: University of Minnesota Press.
Wülker, W. (1985). Changes in Behaviour, Flight Tone and Wing Shape in Nematode-Infested Chironomus (Insecta, Diptera). *Parasitology Research*, Vol. 71, Issue 3.

## 8
## The Great Hacking of the Umwelt

Akpan, N. (2015). The Ant, the Butterfly and Their Chemical Warfare with an Oregano Plant. *PBS NewsHour*. Retrieved from http://www.pbs.org/newshour/updates/ant-butterfly-chemical-warfare-oregano-plant/.
Barbero, F. (2009). Acoustical Mimicry in a Predatory Social Parasite of Ants. *Journal of Experimental Biology*, Vol. 212.
Barbero, F. (2009). Queen Ants Make Distinctive Sounds That Are Mimicked by a Butterfly Social Parasite. *Science*, Vol. 323.
Chakra, M., Hilbe, C., and Traulsen, A. (2014). Plastic Behaviors in Hosts Promote the Emergence of Retaliatory Parasites. *Scientific Reports*, Vol. 4.

Chakra, M., Hilbe, C., and Traulsen, A. (2016). Coevolutionary Interactions Between Farmers and Mafia Induce Host Acceptance of Avian Brood Parasites. *Royal Society Open Science,* Vol. 3.

Cherry, M., and Bennett, A. (2001). Egg Colour Matching in an African Cuckoo, as Revealed by Ultraviolet-Visible Reflectance Spectrophotometry. *Proceedings of the Royal Society B,* Issue 268.

Davies, N. (2000). *Cuckoos, Cowbirds and Other Cheats.* London: T & AD Poyser.

Davies, N. (2011). Cuckoo Adaptations: Trickery and Tuning. *Journal of Zoology,* Vol. 284.

Greenfieldboyce, N. (2016). How Wild Birds Team Up with Humans to Guide Them to Honey. *NPR.* Retrieved from http://www.npr.org/sections/thesalt/2016/07/21/486471339/how-wild-birds-team-up-with-humans-to-guide-them-to-honey.

Harder, B. (2002). Ants Pawns in Battle of Wasps, Butterflies. *National Geographic.* Retrieved from http://news.nationalgeographic.com/news/2002/05/0530_020530_ants.html.

Hojo, M., Pierce, N., and Tsuji, K. (2015). Lycaenid Caterpillar Secretions Manipulate Attendant Ant Behavior. *Current Biology,* Vol. 25, Issue 17.

Holldobler, B., and Wilson, E. (1990). *The Ants.* Cambridge, MA: Belknap Press.

Lhomme, P., et al. (2012). Born in an Alien Nest: How Do Social Parasite Male Offspring Escape from Host Aggression? *PLoS ONE,* Vol. 7, Issue 9.

Lhomme, P., et al. (2015). A Scent Shield to Survive: Identification of the Repellent Compounds Secreted by the Male Offspring of the Cuckoo Bumblebee *Bombus vestalis. The Netherlands Entomological Society,* Vol. 157.

Patricelli, D., et al. (2015). Plant Defences Against Ants Provide a Pathway to Social Parasitism in Butterflies. *Proceedings of the Royal Society B,* 10.1098/rspb.2015.1111.

Sala, M., et al. (2014). Variation in Butterfly Larval Acoustics as a Strategy to Infiltrate and Exploit Host Ant Colony Resources. *PLoS ONE,* Vol. 9, Issue 4.

Sekar, S. (2015). Caterpillar Drugs Ants to Turn Them into Zombie Bodyguards. *New Scientist.* Retrieved from https://www.newscientist.com/article/dn27982-caterpillar-drugs-ants-to-turn-them-into-zombie-bodyguards/.

Starling, M., et al. (2006). Cryptic Gentes Revealed in Pallid Cuckoos *Cuculus pallidus* Using Reflectance Spectrophotometry. *Proceedings of the Royal Society B,* Issue 273.

Thomas, J., et al. (2002). Parasitoid Secretions Provoke Ant Warfare. *Nature,* Vol. 417.

Topoff, H., and Zimmerli, E. (1993). Colony Takeover by a Socially Parasitic Ant, *Polyergus breviceps*: The Role of Chemicals Obtained During Host-Queen Killing. *Animal Behaviour,* Vol. 46, Issue 3.

Van Honk, C., et al. (1981). The Conquest of a *Bombus terrestris* Colony by a *Psithyrus vestalis* Female. *Apidologie,* Vol. 12, Issue 1.

Welbergen, J., and Davies, N. (2011). A Parasite in Wolf's Clothing: Hawk Mimicry Reduces Mobbing of Cuckoos by Hosts. *Behavioral Ecology,* Vol. 22.

Wild, A. (2009). Trophallaxis. *Myrmecos.* Retrieved from http://www.myrmecos.net/2009/10/19/trophallaxis/.

Wilson, E. (1998). *Sociobiology.* Cambridge, MA: Belknap Press.

## 9
## The Brain-Hacked Mouse That Wore a Funny Hat and Destroyed the Notion of Free Will

Barrett, L. (2011). *Beyond the Brain: How Body and Environment Shape Animal and Human Minds.* Princeton, NJ: Princeton University Press.

Coyne, J. (2010). The "Free Will" Experiment. *Why Evolution Is True.* Retrieved from https://whyevolutionistrue.wordpress.com/2010/07/28/the-free-will-experiment/.

Coyne, J. (2013). Yet Another Experiment Showing That Conscious "Decisions" Are Made Unconsciously, and in Advance. *Why Evolution Is True.* Retrieved from https://whyevolutionistrue.wordpress.com/2013/03/25/yet-another-experiment-eroding-free-will/.

Coyne, J. (2015). You Don't Have Free Will. Retrieved from https://www.youtube.com/watch?v=Ca7i-D4ddaw.

Falkner, A., et al. (2016). Hypothalamic Control of Male Aggression-Seeking Behavior. *Nature Neuroscience,* Vol. 19, No. 4.

Gazzaniga, M. (2011). *Who's in Charge? Free Will and the Science of the Brain.* New York: HarperCollins.

Harris, S. (2012). *Free Will.* New York: Free Press.

Harris, S. (2012). Sam Harris on Free Will. Retrieved from https://www.youtube.com/watch?v=pCofmZlC72g.

Keim, B. (2008). Brain Scanners Can See Your Decisions Before You Make Them. *Wired.* Retrieved from https://www.wired.com/2008/04/mind-decision/.

Kennedy, M. (2017). Mechanical Silver Swan That Entranced Mark Twain Lands at Science Museum. *Guardian.* Retrieved from https://www.theguardian.com/artanddesign/2017/feb/02/mechanical-silver-swan-flies-nest-robots-exhibition-science-museum.

Libet, B. (1985). Unconscious Cerebral Initiative and the Role of Conscious Will in Voluntary Action. *Behavioral and Brain Sciences*, Vol. 4.

Soon, C., et al. (2008). Unconscious Determinants of Free Decisions in the Human Brain. *Nature Neuroscience*, Vol. 11.

Soon, C., et al. (2013). Predicting Free Choices for Abstract Intentions. *PNAS*, Vol. 100, No. 15.

Twain, M. (1869). *The Innocents Abroad*. Hartford, CT: American Publishing.

Van Gelder, T. (1995). What Might Cognition Be, if Not Computation? *Journal of Philosophy*, Vol. 92, No. 7.

Wallace, A. (1858). On the Tendency of Varieties to Depart Indefinitely from the Original Type. Retrieved from http://www.rpgroup.caltech.edu/courses/PBoC%20GIST/files_2011/articles/Ternate%201858%20Wallace.pdf.

Woollett, K., and Maguire, E. (2011). Acquiring "the Knowledge" of London's Layout Drives Structural Brain Changes. *Current Biology*, Vol. 11.

Yong, E. (2011). How Acquiring the Knowledge Changes the Brains of London Cab Drivers. *Discover*. Retrieved from http://blogs.discovermagazine.com/notrocketscience/2011/12/08/acquiring-the-knowledge-changes-the-brains-of-london-cab-drivers/.

## 10
## You, the Undead

Abdulai-Saiku, S., and Vyas, A. (2017). Loss of Predator Aversion in Female Rats After *Toxoplasma gondii* Infection Is Not Dependent on Ovarian Steroids. *Brain, Behavior, and Immunity*, 10.1016/j.bbi.2017.04.005.

Acquisto, A. (2017). Mainer Attacked by Rabid Raccoon Drowns It in Puddle. *Bangor Daily News*. Retrieved from http://bangordailynews.com/2017/06/14/news/midcoast/maine-woman-attacked-by-raccoon-drowns-rabid-animal-in-puddle/.

Andhale, S. (2007). Rabies Patient at Nair Hospital Dies. *Mumbai Mirror*. Retrieved from http://mumbaimirror.indiatimes.com/mumbai/other//articleshow/15674027.cms.

Carruthers, V., and Suzuki, Y. (2007). Effects of *Toxoplasma gondii* Infection on the Brain. *Schizophrenia Bulletin*, Vol. 33, Issue 3.

CDC (2017). Chronic Wasting Disease in Animals. Retrieved from https://www.cdc.gov/prions/cwd/cwd-animals.html

Centers for Disease Control and Prevention (n.d.). The Infectious Path of the Rabies Virus. Retrieved from https://www.cdc.gov/rabies/specific_groups/doctors/transmission.html.

Chaudhry, S., Gad, N., and Koren, G. (2014). Toxoplasmosis and Pregnancy. *Canadian Family Physician*, Vol. 60.

Cook, T., et al. (2015). "Latent" Infection with *Toxoplasma gondii*: Association with Trait Aggression and Impulsivity in Healthy Adults. *Journal of Psychiatric Research*, Vol. 60.

Flegr, J. (2013). Influence of Latent Toxoplasma Infection on Human Personality, Physiology and Morphology: Pros and Cons of the Toxoplasma–Human Model in Studying the Manipulation Hypothesis. *Journal of Experimental Biology*, Issue 216.

Fukiyuki, T., et al. (2004). Novel Insect Picorna-Like Virus Identified in the Brains of Aggressive Worker Honeybees. *Journal of Virology*, Vol. 78, No. 3.

Gluska, S., et al. (2014). Rabies Virus Hijacks and Accelerates the p75NTR Retrograde Axonal Transport Machinery. *PLoS Pathogens*, Vol. 10, Issue 8.

Hattwick, M. (1972). Recovery from Rabies: A Case Report. *Annals of Internal Medicine*, Vol. 76.

Intensive Treatment Helped Ohio Man Survive Childhood Bat Bite (1995). *Washington Post*. Nov. 7.

Ling, V., et al. (2011). *Toxoplasma gondii* Seropositivity and Suicide Rates in Women. *Journal of Nervous and Mental Disease*, Vol. 199, Issue 7.

Mariani, M. (2015). The Tragic, Forgotten History of Zombies. *Atlantic*. Retrieved from https://www.theatlantic.com/entertainment/archive/2015/10/how-america-erased-the-tragic-history-of-the-zombie/412264/.

Matheson, R. (1954). *I Am Legend*. New York: RosettaBooks.

McConkey, G., et al. (2013). *Toxoplasma gondii* Infection and Behaviour—Location, Location, Location? *Journal of Experimental Biology*, Vol. 216.

Pederson, M., et al. (2012). *Toxoplasma gondii* Infection and Self-Directed Violence in Mothers. *Archives of General Psychiatry*, Vol. 69, Issue 11.

Rettner, R. (2018). Could 'Zombie Deer' Disease Spread to Humans? *LiveScience*. Retrieved from https://www.livescience.com/61504-chronic-wasting-disease-spread-humans.html.

*Scientific American*. What Is a Prion? Retrieved from https://www.scientificamerican.com/article/what-is-a-prion-specifica/

Torrey, E., et al. (2007). Antibodies to *Toxoplasma gondii* in Patients with Schizophrenia: A Meta-Analysis. *Schizophrenia Bulletin*, Vol. 33, Issue 3.

Torrey, E., and Yolken, R. (2003). *Toxoplasma gondii* and Schizophrenia. *Emerging Infectious Diseases*, Vol. 9, Issue 11.

Vyas, A. (2013). Parasite-Augmented Mate Choice and Reduction in Innate Fear in Rats Infected by *Toxoplasma gondii*. *Journal of Experimental Biology*, Vol. 216.

Vyas, A. (2015). Mechanisms of Host Behavioral Change in *Toxoplasma gondii* Rodent Association. *PLoS Pathogens*, Vol. 11, Issue 7.

Wasik, B., and Murphy, M. (2012). *Rabid: A Cultural History of the World's Most Diabolical Virus*. New York: Penguin.

Webster, J., et al. (2013). *Toxoplasma gondii* Infection, from Predation to Schizophrenia: Can Animal Behaviour Help Us Understand Human Behaviour? *Journal of Experimental Biology*, Vol. 213.

Yolken, R., Dickerson, F., and Torrey, E. (2009). *Toxoplasma* and Schizophrenia. *Parasite Immunology*, Vol. 31.